IET POWER AND ENERGY SERIES 63

Energy Storage for Power Systems

Other volumes in this series:

Volume 1	**Power circuit breaker theory and design** C.H. Flurscheim (Editor)
Volume 4	**Industrial microwave heating** A.C. Metaxas and R.J. Meredith
Volume 7	**Insulators for high voltages** J.S.T. Looms
Volume 8	**Variable frequency AC motor drive systems** D. Finney
Volume 10	**SF6 switchgear** H.M. Ryan and G.R. Jones
Volume 11	**Conduction and induction heating** E.J. Davies
Volume 13	**Statistical techniques for high voltage engineering** W. Hauschild and W. Mosch
Volume 14	**Uninterruptible power supplies** J. Platts and J.D. St Aubyn (Editors)
Volume 15	**Digital protection for power systems** A.T. Johns and S.K. Salman
Volume 16	**Electricity economics and planning** T.W. Berrie
Volume 18	**Vacuum switchgear** A. Greenwood
Volume 19	**Electrical safety: a guide to causes and prevention of hazards** J. Maxwell Adams
Volume 21	**Electricity distribution network design, 2nd edition** E. Lakervi and E.J. Holmes
Volume 22	**Artificial intelligence techniques in power systems** K. Warwick, A.O. Ekwue and R. Aggarwal (Editors)
Volume 24	**Power system commissioning and maintenance practice** K. Harker
Volume 25	**Engineers' handbook of industrial microwave heating** R.J. Meredith
Volume 26	**Small electric motors** H. Moczala *et al.*
Volume 27	**AC-DC power system analysis** J. Arrillaga and B.C. Smith
Volume 29	**High voltage direct current transmission, 2nd edition** J. Arrillaga
Volume 30	**Flexible AC transmission systems (FACTS)** Y.-H. Song (Editor)
Volume 31	**Embedded generation** N. Jenkins *et al.*
Volume 32	**High voltage engineering and testing, 2nd edition** H.M. Ryan (Editor)
Volume 33	**Overvoltage protection of low-voltage systems, revised edition** P. Hasse
Volume 34	**The lightning flash** V. Cooray
Volume 36	**Voltage quality in electrical power systems** J. Schlabbach *et al.*
Volume 37	**Electrical steels for rotating machines** P. Beckley
Volume 38	**The electric car: development and future of battery, hybrid and fuel-cell cars** M. Westbrook
Volume 39	**Power systems electromagnetic transients simulation** J. Arrillaga and N. Watson
Volume 40	**Advances in high voltage engineering** M. Haddad and D. Warne
Volume 41	**Electrical operation of electrostatic precipitators** K. Parker
Volume 43	**Thermal power plant simulation and control** D. Flynn
Volume 44	**Economic evaluation of projects in the electricity supply industry** H. Khatib
Volume 45	**Propulsion systems for hybrid vehicles** J. Miller
Volume 46	**Distribution switchgear** S. Stewart
Volume 47	**Protection of electricity distribution networks, 2nd edition** J. Gers and E. Holmes
Volume 48	**Wood pole overhead lines** B. Wareing
Volume 49	**Electric fuses, 3rd edition** A. Wright and G. Newbery
Volume 50	**Wind power integration: connection and system operational aspects** B. Fox *et al.*
Volume 51	**Short circuit currents** J. Schlabbach
Volume 52	**Nuclear power** J. Wood
Volume 53	**Condition assessment of high voltage insulation in power system equipment** R.E. James and Q. Su
Volume 55	**Local energy: distributed generation of heat and power** J. Wood
Volume 56	**Condition monitoring of rotating electrical machines** P. Tavner, L. Ran, J. Penman and H. Sedding
Volume 57	**The control techniques drives and controls handbook, 2nd edition** B. Drury
Volume 58	**Lightning protection** V. Cooray (Editor)

Energy Storage for Power Systems

2nd Edition

A.G. Ter-Gazarian

The Institution of Engineering and Technology

Published by The Institution of Engineering and Technology, London, United Kingdom

The Institution of Engineering and Technology is registered as a Charity in England & Wales (no. 211014) and Scotland (no. SC038698).

First published 1994 (0 86341 264 5)
Second edition 2011

The Institution of Engineering and Technology
Michael Faraday House
Six Hills Way, Stevenage
Herts, SG1 2AY, United Kingdom

www.theiet.org

British Library Cataloguing in Publication Data
A catalogue record for this product is available from the British Library

ISBN 978-1-84919-219-4 (paperback)
ISBN 978-1-84919-220-0 (PDF)

Typeset in India by MPS Ltd, a Macmillan Company
Printed in the UK by Berforts Group Ltd., Stevenage

Contents

Preface

The last 50–60 years have seen many changes in electricity supply industry. Since the 1950s, the industry has witnessed successive plant build programmes of nuclear-, coal- and oil-fired stations followed by combined cycle gas stations. The result was the development of conventional stations of various sizes and capabilities to provide a continuous, reliable and affordable supply of electricity. Their operation has been in a centrally organised and controlled market with high-voltage grids delivering efficiencies of scale and security in the use of resources.

Since the 1990s, additional changes have occurred. These have been brought about by the liberalisation of electricity supply, the focus on the environmental effects of electricity generation coupled with the imposition of associated emission controls on existing and new thermal plant and, more recently, the setting of national targets for the exploitation of renewable energy. There is no reason to suppose that the next 50 years will not see further fundamental changes arising from different economic, political and technology developments. One such change that may be anticipated, brought about especially by the growth of generation from renewable energy, will be the development and use of energy storage.

The supply of electrical energy cannot be considered alone, however, the dual power *and* energy supply requirements have to be considered together. The main drawback of renewables is that while able to supply with reasonable assurance a certain amount of energy over a year, most are intermittent (*'variable'* is a better description), some randomly so and thus unable to supply power on demand. For this reason, they are limited in their contributions to the security of power supply. Such variations in power output may cause problems for a power system if the renewable resources penetrate the system on a large scale. Matching renewable generation availability to demand in an electricity supply system requires a more detailed examination of this aspect, namely that concerning the interaction of random intermittency of supply with security, bearing in mind that security of electricity supply requires continuity of power delivery, not energy. It is important to note, however, that renewable generation characteristics tend to be system specific and depend on the physical geography and location of the system.

Power supply variability can be of both a short- and long-term nature, and it is the short-term variability that is more relevant to the use of energy storage. In the case of long-term variability where there are low/zero contributions from intermittent generation, as can occur over several days when a large anticyclone weather system covers much of a total wind fleet, for example, security of supply probably depends on having an installed conventional plant capacity margin measured with respect to peak demand as at present.

In the short term, daily variations of power output can be compensated for by existing conventional plant to a certain degree, although this would entail extra costs. These arise from less efficient part-loading with heavy cycling and rapid load following taking its toll by damaging plant equipment, thus requiring far larger maintenance costs. Alternatively, or in addition, large quantities of rapid response plant could be employed such as open cycle gas turbines burning kerosene (or gas if available) or diesel generators, options that are likewise economically inefficient and polluting. In such circumstances, the use of energy storage capacity merits consideration both technically and economically, especially when at very high penetration of wind, for example, the system costs will rise sharply, because wind provides a greater amount of electricity without necessarily providing capacity to help meet peak demand.

Energy storage possibilities linked to the more efficient production and use of electricity exist in a rich variety of options including chemical, biological, electrochemical, electrical, mechanical and thermal alternatives as described in this book by Dr Ter-Gazarian. New potentially disruptive technologies are now appearing to offer alternative future solutions to meet energy market needs at various levels from the small-scale consumer to utility-scale energy storage.

Disregarding chemical fuels (including hydrogen) that have become the dominant form of energy storage, both in electrical generation and in energy transportation, foremost present developments include:

- *Batteries*: Since about the middle of the first decade of the twenty-first century newer battery technologies have been developed. These can now provide significant utility-scale load-levelling capabilities and, with the introduction of lithium-ion battery technology to electric cars, a significant improvement in performance is leading to the large-scale introduction of electric and hybrid cars to consumers over the next decade.
- *Grid energy storage*: A proposed variant of grid energy storage is called a vehicle-to-grid energy storage system, where modern electric vehicles that are plugged into the energy grid can release the stored electrical energy in their batteries back into the grid when needed.
- *Fuel cells*: Any device powered by electricity can be powered by a fuel cell, but in all of the possible markets the hydrogen supply poses significant problems. The three markets where fuel cells will find application are categorised as stationary, automotive and portable. The stationary power market includes low-power single-family homes; medium-power uses in commercial premises, hospitals and hotels; and high-power uses in multi-megawatt industrial plants. The major technology for transportation purposes has emerged as the low-temperature solid polymer fuel cell (SPFC) and, in particular, the proton exchange membrane fuel cell (PEMFC). The portable power market includes devices requiring from milliwatts to several hundred watts such as consumer electronics, laptop computers to military equipment.

- *Regenerative fuel cell*: Another development is based on fuel cell technology, but that could also be classed as a battery inasmuch as the charge capacity is limited by the quantities of the chemical constituents.
- *Solar PV*: As a renewable source of energy, solar PV has many advantages over fossil fuel electricity generation. It is eminently low carbon and, because energy is locally generated, no distribution losses are associated with energy being transported through the national grid. High cost is still the main barrier hindering solar PV development however, so if scientists can create reliable, low-cost solar cells, that is, cheaper, organic PV products, this could make a significant contribution to meeting growing energy needs and cutting carbon emissions.
- *Large-scale solar power plants*: Insulated containers store the hot molten salt solution heated by the Sun rays and these are then used to create steam that is fed to turbines to generate electricity. Recent developments are seen in Spain and United States, with possible further installations in Israel, Egypt, France, Australia, Algeria and South Africa.
- *Pumped hydro storage*: This is the largest utility energy storage method in the world.
- *Marine energy*: Various options for barrages or tidal lagoons are under consideration, especially in the Severn Estuary in the United Kingdom.
- *Compressed air*: These schemes use compressed air generated by using renewable power to pump air into underground caverns or aquifers, then releasing it to augment natural gas-fired turbines when power demand peaks. After some years, with only two of these projects worldwide – one in Alabama and one in Germany – new facilities are being constructed in the United States.
- *Wind bags*: This is a variation on compressed air storage whereby flexible wind bags are being developed in the United Kingdom to store excess energy from wind turbines as compressed air under the sea.
- *Underground heat storage*: Projects are underway to investigate how geological formations could be used to store large quantities of heat generated by power stations lying in close geographical proximity to those formations. At the same time, it is intended to look at localised energy storage in 'gravel beds' as a way to store heat on a smaller scale. The heat would be stored and pumped into homes and businesses using water, in much the same way as a central heating system operates.
- *Thermal energy storage*: Thermal energy storage has been employed since the 1980s to meet the demand for air conditioning using ice storage for cooling, the ice being produced with off-peak power and employed during peak demand hours. In 2009, it was used in over 3 300 buildings in over 35 countries.
- *Flywheels*: A recent full-scale 20 MW flywheel energy storage plant is being used as a backup power supply for emergencies as well providing assistance in system frequency regulation, that is, supporting quality of supply, a problem of considerable importance.

The present population of the world is doubling about every 35 years while the demand for energy is rising even more rapidly, increasing by 5% per year, equivalent to doubling about every 14 years, and placing increasing demands on the available hydrocarbon sources. In such circumstances, the development of sustainable energy supplies is essential for the long-term future, apart from any environmental considerations. The uncertainties increase substantially, but assuming broadly sufficient MW capacity is provided, the key issue will become one of balancing power supply and demand continuously, noting the large amount of variable output renewables on the system and what is likely to be a rapidly growing level of time-movable demand enabled by smart grids. Renewables will thus have a substantial role, but the use of energy storage facilities enhancing the value of renewables and transmission facilities by providing capacity support will also be significant.

M.A. Laughton
22 December 2010

Acknowledgement

This book would have never been written without Dr D.T. Swift-Hook's encouragement. Chapter 17, in particular, has been written as a result of valuable discussions with him and is partly based on his work. The author also wishes to express gratitude to his wife Mrs Olga Ter-Gazarian for her patience and care in typing the initial manuscript, and to Ms Kathy Abbott for her kind hospitality and assistance during his essential visits to London.

Introduction

Energy conversion: from primary sources to consumers

Our civilisation depends on energy. Technical progress and development from prehistoric times were connected with the quantity of energy used. This can be illustrated with the help of Figure I.1. A sharp increase in energy consumption started after the Second World War. More than two-thirds of the overall quantity of 900 000–950 000 TWh of energy has been consumed during the past 50 years and 90% of this energy is not renewable. It is not surprising, therefore, that serious interest in primary energy sources has arisen in recent decades: energy demand is growing, but conventional energy sources are limited and not located everywhere.

Energy occurs in a variety of types depending on the nature of the interaction used in transforming it from one form to another. There are only four types of interaction: gravitational, weak nuclear, electromagnetic and strong nuclear, but their interactions with the constituents of the universe, namely baryons, leptons and photons, produce a large number of different forms of energy. The principal types of energy are potential, kinetic, electrical and mass or radiation. The various forms of energy commonly encountered are listed in Table I.1.

The global energy picture, depicted schematically in Figures I.2 and I.3, shows that practically all our energy sources are connected in one way or another to the Sun. Near the equator, at the outer atmosphere, our planet receives from the Sun about 1360 J of energy per second over every square metre, that is, 1.36 kW/m^2. At the sea level, we receive approximately 1 kW/m^2, or just over 1 horsepower per square yard.

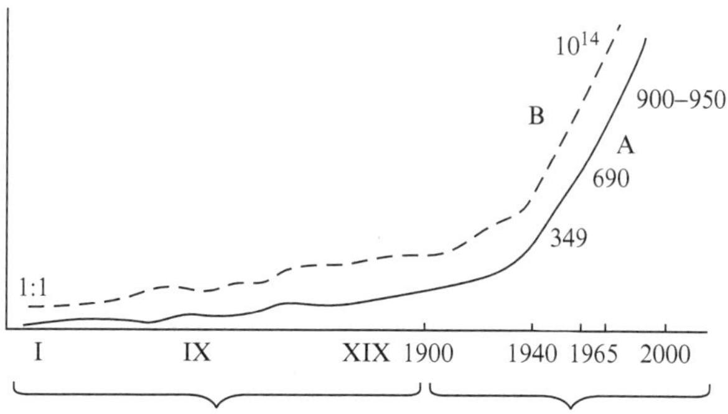

Figure I.1 Energy consumed by humankind (A, 10^6 GW) and stored information (B, bt) from prehistoric times to the present

Table I.1 Physical forms of energy

Type	Form	Potential	Quantity	Storable or not	Example
Gravitational	Altitude	Gravitation potential, g_h	Mass, m	Storable	Dam
Kinetic	Velocity, v	$v/2$	Mass, m	Storable	Flywheel, Bullet
Spatial	Pressure	Pressure, P	Volume, V	Storable	Compressed gas
Thermal	Heating	Temperature, t	Entropy, S	Storable	Hot water
Chemical	Electron charge	Chemical potential, G	Number of moles, n	Storable	Chemical battery
Electric	Electron charge	Voltage, U	Electric charge	Storable	Capacitor
Dielectric		Electric field, E	Dielectric polarisation, P	Not storable	Polarisation
Magnetic	Electron spin	Magnetic field, H	Magnetisation, M	Not storable	Magnetisation
Electromagnetic	Moving charge	Self-inductance voltage, $L\ dI/dt$	Electric current	Storable	Magnetic coil
Weak nuclear	Mass change		Mass	Not storable	Luminous paint
Strong nuclear	Mass change		Mass	Not storable	Nuclear reactor, stars
Radiation	Photon			Not storable	

But we are not consuming the Sun's energy directly – most, if not all, of our sources use the Sun's energy stored in different ways.

There are plenty of so-called natural energy storage media – organic fuel (wood, coal, oil), water evaporation and wind. Energy from the Sun has been accumulated for billions of years in organic fuels, for years owing to evaporation from rivers and for seasons in the wind.

In transforming energy from one form to another, there is sometimes a mechanism for storage that enables a reserve of energy to be established for subsequent use and at a rate that is not dependent on the initial rate of transformation. This twofold decoupling in time of the energy transformation processes has far-reaching consequences, since it is exactly in that way that Nature allows us to concentrate energy and to consume it some time after its generation.

Generally speaking, energy flow from a primary source is not constant but depends on season, time of day and weather conditions. Energy demand is not constant either; it depends on the same circumstances but mostly in reverse. So there needs to be a mediator between the source of energy and its consumer. This mediator is energy storage, which, in one way or another, plays a role in all natural and man-made processes.

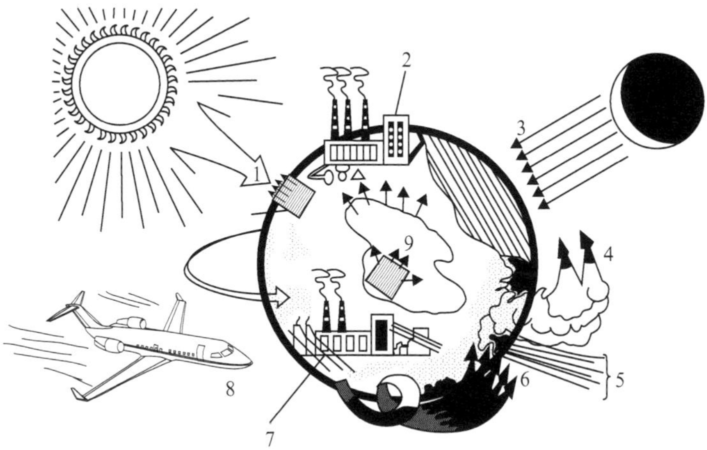

1. Solar radiation 100% ($E = 7.5 \times 10^{14}$ MWh/year)
2. Power installations 0.01%
3. Tidal power 0.05%
4. Wind 0.035%
5. Evaporation 0.0005%
6. Hurricane 0.04%
7. Power stations 0.0015%
8. Aircraft 0.0002%
9. Direct reflection 30%

Figure I.2 Comparison of artificial and natural power flows

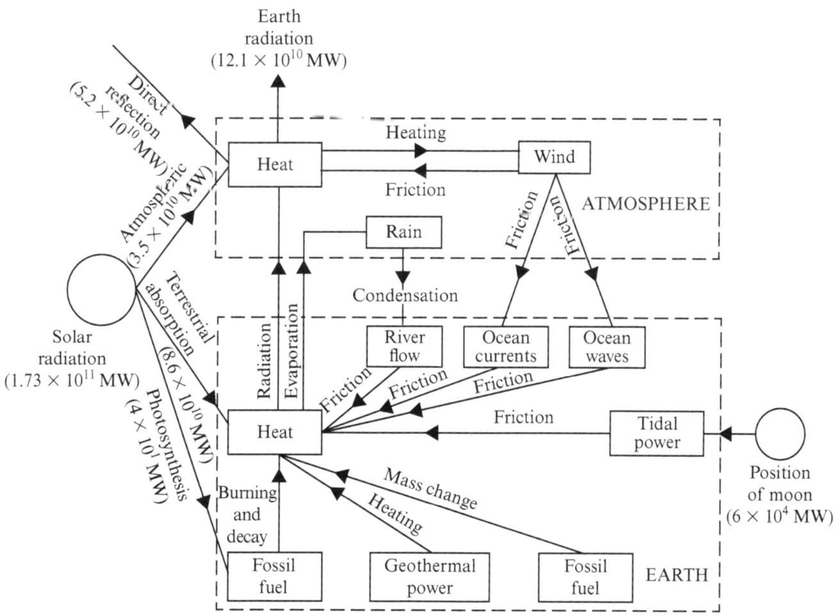

Figure I.3 The global energy picture

Energy storage is an essential part of any physical process, for without storage all events would occur simultaneously; it is an essential enabling technology in the management of energy.

Historically, the energy storage problem was solved by piling lumps of wood together, or by damming springs to provide a working head for a waterwheel. Later, a more concentrated form of fuel – coal – became the most important energy store. Today, we are accustomed to the filled petrol tank as a most convenient form of energy storage. Oil-based fuels offer easy usage, ready availability and relatively low price. Technically, the storage of oil is not difficult and the storage time simply depends on when the tap of the tank is turned off. Storage can be maintained without any losses for any length of time and the energy density is high. Oil can be used as a source for electrical power and heat, for transportation and also for stationary applications. Comparison of the different types of natural storage is given in Table I.2.

Table I.2 Fuels as natural energy storage

Fuel type	**Density** 10^3 **kg/m**^{3}	**Energy density**		**Boiling point** °**C**
		10^7 **J/kg**	10^{10} **J/m**	
Wood	0.48	1.5	0.72	N/A
Coal	1.54	2.93	4.5	N/A
Petrol	0.786	4.12	3.24	127
Propane	0.536	4.7	2.52	–42.2
Methane	4.2	0.5	2.1	–161.3
Uranium		8.2×10^3		N/A

All the primary fuel- and energy-producing industries, including electricity generation, make significant use of energy storage for efficient management of their systems in both the initial extraction of primary fuel and the subsequent distribution to consumers. Stocks of primary fuel are used to provide security against interruption in production, to permit stability in pricing of energy delivered to the consumer and to match the varying load demands to a nearly constant primary fuel production rate. Examples are coal mountains, oil tank farms and compressed natural gas storage in the underground caverns. Storage is again used to maximise the efficiency of major distribution systems, namely coal depots, garages and gas cylinders including line packing.

Renewable energy sources have attracted considerable interest. However, most renewable sources do not provide a constant energy supply and, unlike fuels, cannot be directly stored, so they either require secondary storage systems or have to be used in power systems with sufficient reserve capacities installed on conventional power plants.

Electrical power systems are the most flexible and convenient carriers between energy producers and consumers. Electrical power systems have some exceptional properties. Some of them are listed below:

- Mechanical–electrical converters are robust, efficient and easily controllable.
- Electrical transmission lines are very efficient and can transport energy over long distances reliably and economically.
- Interconnected electrical power systems are widespread in most of developed countries. Therefore, 'in principle' it is easy to feed the converted energy from intermittent sources into the grid.
- It is economically advantageous to offer the energy from intermittent sources in electrical form since it costs more per kWh compared with energy in a solid or liquid fuel. This is because it is 'high-grade' energy and, in its generation from fuels, the Carnot cycle imposes a low efficiency of conversion.

The main disadvantage of electricity as an energy carrier is the impossibility of storing it in sufficient quantities. The problem of energy storage becomes more and more crucial with power system development when the capacity of the units is increasing.

It is possible to improve this disadvantage by the use of intermediate devices on the power system, which can partly or completely separate the processes of energy generation and consumption. We will call these 'secondary energy storage'.

Secondary energy storage (SES) is an installation specially designed to accept energy generated by the power system, convert it into a form suitable for storage, keep it for a certain time and return as much of the energy as possible back to the power system, converting it into a form required by the consumer.

A slightly simplified rule states that storing heat saves energy while storing electricity saves capital investment. Since the energy industry on a global basis faces a resource crisis, due to both primary energy and capital investment, there are good reasons for investigating the storage of low-quality as well as high-quality energy.

The use of energy storage in the form of chemical batteries for bulk power supply is not a new concept: they were the essential part of many electricity utilities at the beginning of this century. The earliest application of electric storage batteries was in German town, Pennsylvania, in 1890. At that time, electric power systems were primarily direct current (DC) systems and operated isolated from each other. Electric storage batteries were used to provide peak electric energy requirements and emergency capacity when dynamos, which generated direct current, were not in operation. With the expansion of DC electricity systems, the use of batteries became widespread, but development of a competitive alternating current (AC) electricity system changed the trend dramatically. As AC systems evolved and became widely accepted, owing to advances in equipment and systems design that increased their reliability, the use of batteries was discontinued.

As the magnitude of energy transformation increases in time with increasing industrialisation, the need for efficient energy storage and recovery also grows. The large-scale transformation of chemical energy stored in fossil fuels in recent years has emphasised the need for conservation of energy, which can often be achieved by introducing some form of energy storage into the system.

Even though the cycling of energy through storage causes inevitable losses, the overall economic benefits are significant. These benefits are derived from two principal areas: reduced primary fuel consumption from increased efficiency of operation including conservation and development of new equipment related to, or dependent on, storage techniques.

Basically, energy is stored at times when the available means of generation exceed demand and is returned when demand exceeds generation.

Of the four forms in which energy is consumed, that is, mechanical, thermal, light and electrical, in practice only thermal energy can be stored by consumers connected to the electricity power system. Other forms of energy, which are transmitted at the consumer's disposal only by means of electrical energy, can be stored locally, mostly in the form of electrical energy but at much higher cost than storage at the supply side.

In fact, apart from the self-evident effect of scale, a producer can use storage to level a load demand in which the effect of time diversity of the peaks of the various consumers is already completely utilised, whereas this cannot be done by a single user. In other words, if each consumer used his own storage means for levelling his own load consumption, it is likely that many storage devices would be in the generation phase while others would be at standby or even in the storage phase, meaning that overall utilisation of the storage capacity would be low. Siting and control problems would also arise.

Among consumers, only isolated users who cannot be connected to the electricity power system, owing to distance, and electric vehicles, owing to their forced autonomy, can conveniently store energy themselves, as they already do for energy generation.

The installation of energy storage units on electricity utility networks will permit the utilities to store energy generated at night by coal-burning or nuclear baseload plants, and release the stored energy to the network during the day when the demand is highest. This reduces the need for gas- or oil-burning turbines.

Large-scale utilisation of renewable energy sources such as the Sun and the wind depends on storage facilities, since these sources are variable throughout the day and through the year.

Storage at the supply side is able to play roles different from the simple ones described earlier because storage means usually have characteristics very different from those of generation means. Therefore, energy storage can perform other functions, complementary to those of the generating equipment and very important from both technical and economical points of view. These functions are considered in Part 1 of the book. Part 2 of this book expands upon the different storage techniques.

The problem of selecting the most appropriate composition of a mixed generating and storage structure appears immediately to be the most important problem for the power system engineer. Therefore, Part 3 of this book is devoted to this extremely complex problem that can be solved provided the information on power system requirements for energy storage, as well as storage-equipment technical and economical parameters, are known.

Part I

The use of energy storage

Chapter 1
Trends in power system development

1.1 Demand side characteristics

The demand side of the power system is made up of consumers in three categories: industrial, domestic and commercial (the last including public lighting). Data given in Table 1.1 show the energy consumed by these groups in some European countries and in Russia. Each group has a considerable influence on the total energy consumed by the demand side of the power system but each has its own peculiarities. Maximal consumption in the domestic sector, for example, occurs during morning and evening hours and weekends when people, being at home, use most of their electrical devices.

Table 1.2 gives an illustration of the power requirements for different domestic appliances, while curves in Figure 1.1 show how domestic consumption is shared between the different components.

Public lighting requires power only during the evening and in a reduced quantity during night hours. Domestic consumption is at a minimum during the daytime and also at night. On the contrary, commercial consumption reaches its maximum during the daytime, particularly at lunch hours, and at the end of a working day. Industrial consumption is more stable than domestic and commercial consumption because of the possibility of organising work on several shifts, although not every industrial consumer can operate shifts, so maximal consumption occurs during the daytime with a slight decrease at lunchtime (the 'daytime hollow').

In Western countries, which have considerably higher living standards, domestic and commercial demands represent a significant part of overall consumption, which therefore has visible morning and evening peaks and night-time troughs of demand.

Table 1.1 Sharing of electricity consumption by sectors

	Russia	Germany	France	UK	Italy	Spain
Industry	0.646	0.475	0.442	0.374	0.542	0.560
Transport	0.089	0.031	0.027	0.017	0.027	0.029
Domestic	0.078	0.264	0.307	0.350	0.260	0.226
Commerce, agriculture and public lighting	0.187	0.230	0.224	0.260	0.179	0.185

*Table 1.2 Average power requirements for various domestic appliances**

Domestic appliance	**Required power W**	**Monthly energy consumption kWh**
Air conditioner	1300	105
Boiler	1357	8
Fan (attic)	375	26
Fan (window)	190	12
Heat lamp (infrared)	250	1
Heater (radiant)	1300	13
Water heater (standard)	3000	340
Refrigerator	235	38
Refrigerator-freezer	330	30
Refrigerator-freezer (frostless)	425	90
Clothes dryer	4800	80
Dishwasher	1200	28
Vacuum cleaner	540	3
Washing machine (automatic)	375	5
Washing machine (non-automatic)	280	4
Coffee-maker	850	8
Deep-fat fryer	1380	6
Food blender	290	1
Food mixer	125	1
Fruit juicer	100	0.5
Frying pan	1170	16
Grill (sandwich)	1050	2.5
Oil burner or stoker	260	31
Radio	6	0.5
Television	300	37
Toaster	1100	3
Electric blanket	170	12
Hair dryer	300	0.5
Hot plate	1250	8
Iron	1050	11
Roaster	1345	17
Sewing machine	75	1
Shaver	15	0.2
Sun lamp	290	1

*McGuiden, D.: *Small Scale Wind Power* (Prism Press, UK, 1978).

In countries such as China and India, where domestic consumption is still limited owing to the lack of relevant appliances, and most industrial enterprises work on a three-shift basis with a sliding day off, overall consumption is quite stable with rather small fluctuations. The load demand factor in Chinese and Indian utilities is always higher than that in European utilities. However, with the living standards rising, the load demand factor lowers in these countries.

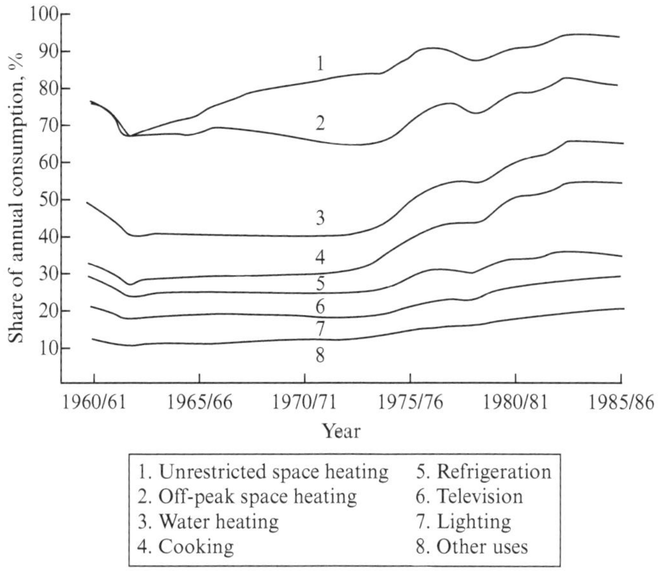

Figure 1.1 Estimated component shares of domestic electricity consumption

The effect of domestic consumption becomes more obvious in big towns and in a number of non-industrial developing countries.

The shape of the daily consumption diagram, its weekly and seasonal diversification and their statistical features depend on the behaviour of the different consumer groups. Different patterns occur in different areas not only in a continent-wide country but also in different regions of a smaller one: such variation depends on differences in the structure of the load as mentioned above, but it can also be strongly influenced by different climatic characteristics. The maximum power consumption in California, for example, occurs during the summer owing to air conditioning requirements, while in Canada seasonal peak demand occurs in the winter owing to heating requirements. In UK power systems, as well as in those of other European countries, the peak demand normally occurs in the afternoon of a weekday in winter. System load factors vary but they are within a range 0.4–0.6.

Figure 1.2 illustrates the main features of the power system's demand side, showing a typical west European country's winter weekday load curve. At other times of the year the peak is more likely to occur in the morning or, during weekends, at lunchtime.

A survey of the various load conditions shows that in essence the majority of problems of peak coverage occur at times quite different from those of maximum load during the winter weeks. The most difficult to meet are load peaks in the daily load curve, particularly if they are relatively high, irrespective of the absolute load level, in comparison to the load of the preceding and subsequent minutes and hours. This is the case in winter as well as in summer or during transition periods, but also

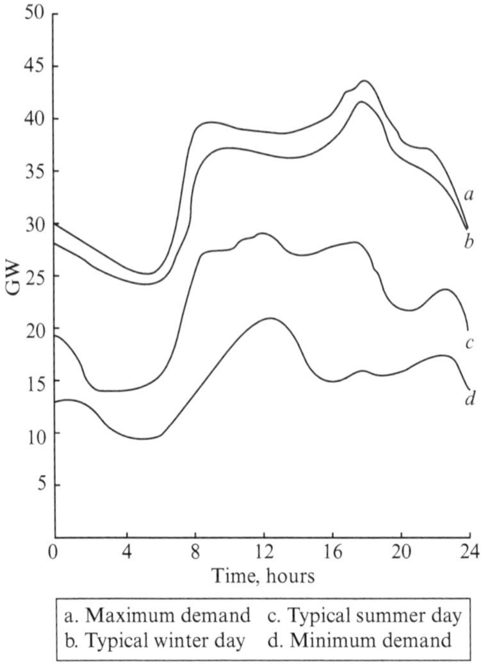

Figure 1.2 Summer and winter demand on a typical system, including days of maximum and minimum demand

on Saturdays and Sundays, or at times where the peaks are not as high as the respective daily peaks. The massive switch-on of storage heating units after the transition from the high-load to the off-peak load period, television viewing on special occasions, strikes and unusual changes in the weather can be mentioned amongst the causes of such variability.

The characteristics of the daily load consumption curves, as is clear from Figure 1.2, are not fixed, but their evolution depends on the slow modification of the structure of consumers in the aggregate, and on individual consumer behaviour, which in many ways is related to economic and social development.

1.1.1 Storage methods

Figure 1.3 also shows how the shape of the daily load curve depends on domestic, industrial and commercial consumption. The profiles of these demands are fairly stable although there are some possibilities of changing them, using tariff policy, introducing different administrative clock times for different regions or interconnecting remote systems with different clock time owing to their longitude, or involving other load-management provisions. These provisions we will call storage methods.

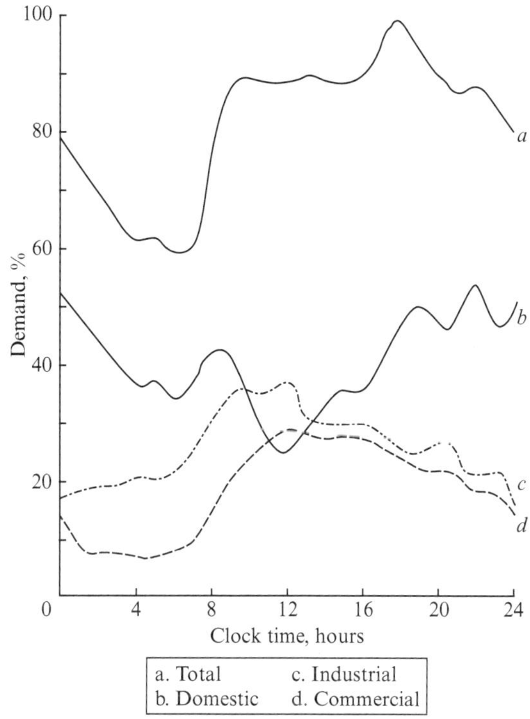

Figure 1.3 Midwinter weekday load curves for average cold spell conditions

A tariff structure, however, cannot be as flexible as might be desirable. Moreover, since it must remain valid for a certain period, it also necessarily includes certain adjustments that partly prejudge the expected result. Furthermore, by its very nature, any tariff structure is difficult to adapt to the uncertain load requirements from the demand side.

The other possible way to change the load curve is advancing the clock time. Between 1968–9 and 1970–1, the United Kingdom tried the experiment of advancing clock time by 1 h in winter months as well as the summer, thus moving from Greenwich Mean Time to a new 'British Standard Time'. The effect of this was to alter the winter weekday load shape, lowering the evening peak and raising the morning peak so as to give a flatter daily plateau.

In the mid-1980s there was an experiment by the Russian power utilities when different districts had different clock times. This experiment was technically successful.

A number of utilities have had contracts for disconnectable loads. Under the current load-management scheme, these loads can be disconnected for up to 60 h in aggregate within the period from 1 October to 31 March for the northern hemisphere countries, which supplements the in-depth effect of tariffs alone.

Typically it is expected that approximately 30 h of disconnection will be required in an average year, usually over 2 h during the evening peak period of the days concerned. The demand that remains after such reduction is called the restricted demand, and it is this demand that must be met by generating plant. Since the early 1960s a number of power utilities have succeeded in increasing the night load substantially by promoting sales of electricity at cheaper night rate tariffs for space and water heating.

There are many remote-control and regulation procedures, which in fact 'manipulate' the load curve: the consumer is able to ascertain the appropriate reaction in response to the signal sent to him by the supply side of the power system and carry it out to the degree economically desirable for himself. This result, of course, implies that the economic signal correctly reflects the real costs, which is a problem. But it is also necessary for the consumer to receive adequate information because his reflexes are not always rational. For example, reducing heat consumption at night, with the laudable intention of effecting savings, is worthwhile only in the case of heating with gas or coal; it does not make sense in correctly insulated, electrically heated premises since at present the increased daytime consumption demand imposed on the least economical power stations will quickly exceed the saving in heating calories, quite apart from the inconvenience caused to the user by such reduction.

1.1.2 Daily load curve structure

A daily load curve is usually divided into three layers: base, intermediate and peak. According to the definition, the base layer is that part of the load curve that does not vary during the day. The intermediate layer is the part of the load curve that varies no more than twice a day. The remaining part of a load curve is considered to be a peak load. As is clear from Figure 1.3, the domestic and commercial parts of the load demand have the main influence on the peak and intermediate load, while industrial load mostly influences the base load, which is growing more slowly than the peak and intermediate loads.

The parameters of the load consumption curves are usually grouped according to their different nature:

- static characteristics related to demand values without reference to the transition from one to another and
- dynamic characteristics related to demand changes, all relevant parameters being identified by their own statistical distribution.

As a function of time the load curve shows several peaks and troughs. Among these are the maximum peak L_{max} and the minimum hollow L_{min}, and their ratio called the minimal load factor

$$\beta = \frac{L_{min}}{L_{max}}$$

has a particular interest.

Other important characteristics of the load curve are as follows:

- range of power variation δL required by the demand side, which is given by

$$\delta L = L_{\max} - L_{\min}$$

- load density factor γ, which is equal to the ratio of daily energy consumption and maximum daily energy consumption, and is given by

$$\gamma = \frac{\left(\int_0^{24} L(t)dt\right)}{24L_{\max}}$$

- time of maximum load $T_{\max}$, which is given by

$$T_{\max} = \frac{\left(\int_0^{24} L(t)dt\right)}{L_{\max} = 24\gamma}$$

- load rise or fall rate, which is given by

$$L_v = \frac{dL}{dt}$$

and is the dynamic characteristic of a demand curve.

It should be pointed out that the largest variation in the amplitude of load demand is not necessarily related to the maximum peak and the highest rate is not necessarily related to the largest variation of a load curve.

The pattern of daily diagrams and all the above-mentioned characteristics are different for various days of the week: working days (separate consideration is generally needed for Monday), Saturday and Sunday (likewise holidays). And for these typical days the pattern can differ according to the season.

Usually the power system has a seasonal pattern with the highest demands occurring either in the winter or in summer months, depending on the geographical conditions. The average daily peak demand in the summer is about two-thirds of the average daily peak demand in winter, but the minimum summer night consumption is only about one-fifth of the winter peak load. It is necessary to have sufficient generating equipment available to meet the peak demand of the year, and this, in turn, is very sensitive to the incidence of spells of particularly cold weather.

For a planning horizon, a forecast of peak demand is made for an average expectation of severe weather, and this is called the average cold spell (ACS) winter peak demand. The ACS demand is approximately 9% higher than it would have been in average winter weather conditions. But the effect of a really severe winter with prolonged spells of low temperature is far worse than even this. In severe winters demand may reach a level even some 10% higher than the ACS demand, but the supply side structure is not planned for that, so the excess demand should be disconnected.

The maximum value of the variable load in European countries generally occurs in the summer months (noon peak); the maximum value of the daily constant load, in general, occurs during the winter months (influence of storage heating).

The night-time trough in the winter diagram rapidly fills up as a result of the introduction of night-storage heating; the peak/trough ratio has decreased since 1965 from 100:35 to approximately 100:70. It has become possible to meet winter demand predominantly by base-load energy. The summer load curve becomes more and more ‘critical’ from an operating and technical standpoint since its shape varies slightly with an almost constant peak/trough ratio of approximately 100:40. The 2-h peaks and the steepness of the load slopes have approximately doubled between 1965 and 1990.

Differences affecting the whole daily diagram, perhaps for several consecutive days, may appear from one year to another as a result of the influence of special economic circumstances, or climatic conditions, on the electrical energy and power demand. Therefore, all the characteristic parameters indicated previously (values of peaks and troughs, variation number, amplitude, rate, etc.) are random variables with a distribution that has to be statistically investigated and represented. Not only the average values, but in many cases the whole distribution and extreme values, with associated probabilities, have to be taken into account.

It is difficult and obviously too cumbersome to represent the daily load pattern as a continuous function of time. A stepwise model is often adopted as an approximation to handle in studies and for calculations, particularly corresponding to the hourly or half-hourly sampling of measurements. It should be kept in mind that such approximations mask the actual rise or fall of load and can give misleading information on their actual rates. Special care should be devoted to collecting information on these rates and to relating them to the stepwise model. The statistical nature of step levels should in any case be considered.

It is obvious that electricity is not the consumer’s only concern. He will seek to modulate all the energy requirements, particularly by requiring one form of energy to serve as auxiliary to another when it is advantageous for him to do so. It is therefore generally advantageous that the auxiliary source should be clearly more economical, in terms of specific investment, than the basic energy source over the whole producer/ultimate-consumer chain. For example, using oil heating as an auxiliary to a heat pump is a step in the right direction, but using electricity as an auxiliary to solar or geothermal heating is probably not the best choice. Other energy systems will undoubtedly have an effect on user load curves in the future. We must therefore consider this problem, which is directly linked (whether favourable or otherwise) to the question of storage and does not come within the purview of the other groups. The threat of unusual demand for electric power during very cold or very dark periods already exists. On the other hand, there may also perhaps be possibilities of encouraging off-peak consumption or leveling out peaks. In this connection, mixed heating systems comprising a heat pump and an oil-fired boiler offer interesting possibilities.

The producer-distributor will try to identify the services expected by the user, will sometimes participate in designing appliances that are satisfactory to both the user and the distributor, and will then envisage the establishment – unless one exists already – of a tariff that encourages use of the appliance and that may benefit all other potential applications, without distinction between them.

1.2 Supply side characteristics

From their earliest days, electricity utilities have been confronted with the contradictory problem of meeting fluctuating demands for power at the lowest possible cost with the reliability required.

The performance of the power system was therefore considered in the context of covering the above-mentioned three-layer load demand curve, which varies according to the previously described parameters, by means of the supply side that is essentially the set of all the installed electrical generators.

From this entire set of installed units, different subsets (depending on, e.g., the availability of water for hydro units, maintenance schedule and forced outages) can be committed to operation by the utility's planning personnel. The sets of committed units actually operating can be different for different load levels; the unit commitment and the load dispatch among the different units should be made taking into account their economic characteristics.

The supply side of the power system has to be provided at power levels with different durations; the peak generation is that related to the power range associated with the shortest durations and therefore with a lower amount of delivered energy. On the other hand, the base-power generation is related to the longest duration. An intermediate or modulation duty can be considered in between.

In relation to planned and forced outages, the maximal installed capacity should have a reserve margin above the maximal load peak in the period considered. Reserve intervention during forced outages should be assured by spare capacity of committed units that are underloaded and by spare capacity of a certain part of the available generating units not called on to cover the load but always ready to go into power supply service.

Some of these units must be able to go into operation immediately or within a few seconds, in order to ensure frequency recovery in the event of a sudden loss of a certain number of generating units, thus preventing massive load cut-off.

A part of the reserve means must be able to take over the load within a few minutes in order to minimise the possibility that a further outage of generating units could lead to inadmissible drops in frequency.

It is convenient to classify the requirement for reserve capacity into the following three categories:

- Immediate or spinning reserve: to cover the initial transient (5–10 s) following plant or infeed (e.g., cross-Channel link) losses.
- Dispatching or operating reserve: to cover despatching allocation of load to generating units' errors on a timescale of several minutes. This implies a

sufficient response by governor action of the regulating plan to system frequency errors.
- Scheduling or ready standby reserve: to cover load and plant changes over the period required to bring new steam plant up to load, which usually takes hours.

In addition to the technical requirements of operation, i.e. rapid operability, good controllability and a certain availability rate, this capacity category also calls for consideration of the economics of demand coverage. Already the low utilisation of 1000 h per year or less makes this necessary. Regard must be paid, however, to the variable costs as well, particularly if peak capacity is supplied by thermal units operating on part load in the medium-load range.

Traditionally, the utilities have responded to load demand by installing combinations of generating plants with different operating and economic characteristics to fulfil the demand side requirements.

Each type of generating unit available for supply side composition has from the technical point of view its own 'natural' characteristics, which should be grouped as follows:

1. static characteristics related to the capability to deliver energy at a required amount of power, and;
2. dynamic characteristics related to the ability to be loaded or unloaded at a required rate.

As far as static characteristics are concerned, the main difference lies between thermal units, which have, in principle, unlimited primary energy available, and hydro units, which have a more or less limited amount of primary energy at their disposal. In the first case, the duration of power output is, in principle, unlimited, but in the second there is a link between power output and its duration.

For thermal units, any output value between the technical minimum and the rated power capacity can be obtained; some minor restriction may derive from the temperature of cooling fluids, but an output over the normal rating can be obtained for more or less limited periods by the reduction or the bypass of steam extraction for feedwater heaters. In the same way, steam extraction for district heating can be temporarily reduced, the heat delivery being compensated by sensible heat storage capacities of the heat distribution system; when back pressure turbines are utilised for this purpose, the heat storage capacities make it possible to increase the heat production and related electric power during limited periods.

With regard to hydro units, the head variation is reflected in the maximum capacity, and the constraint on their capability imposed by the availability of water has a stochastic nature modified by the policy of reservoir exploitation.

The transient characteristics of different generating units during their operation are expressed by the rate of change of output; this rate can have different values in loading and unloading and for different amplitudes of variation ranges and different starting output levels.

Other very important characteristics concern the fact that generating units cannot be kept in operation continuously. For maintenance and other outages, which can be planned in advance, the relevant characteristic, the maintenance duration or the outage time, is of a deterministic nature. Unplanned outages, on the other hand, caused by faults and which can suddenly reduce the output capacity, can be expressed by random variables and statistical parameters. The starting of unit operation requires consideration of a special transient characteristic – the minimum start-up time – as required in particular for emergency commitment. For thermal units, it may depend on the duration of the preceding rest period.

The associated economic characteristics affect the build-up and operation of generating systems no less than the above-considered technical characteristics of generating units. The presence of a generating unit of a given type, in operation or at rest, involves a fixed cost that is related to the expense of construction, personnel, etc.

Hence, electrical energy is being produced with variable efficiency and at variable cost, depending on both the cost of fuel and the power output.

As far as dynamic characteristics are concerned, the start-up cost of thermal units related to the fuel consumption, including the cost of the heat lost during the preceding shutdown, should be determined and taken into account. Much more difficult is the evaluation of the fatigue cost associated with the different nature and frequency of output changes, in particular start-up and shutdown.

The whole class of dynamic characteristics is only briefly mentioned here to remind us of its importance in the context of generating unit performance: this is the class of characteristics concerning the rotating motion of machines under the action of the torque resulting from imbalance between driving and loading powers. Generally speaking, one group of relevant parameters that may be expressed as time constants relates to the inertia of the rotor, the inertia of the driving fluid in the system feeding the turbine and the thermodynamic behaviour of steam. Another group of parameters is represented by time constants and gain coefficients related to the performances of the unit governor system.

Generally speaking, the desired generating structure may be produced by the following types of power generation installations:

- hydro plants;
- gas turbine plants;
- conventional thermal plants;
- renewable plants; and
- nuclear power plants.

Hydro plants are the oldest and most reliable type of generating equipment. They can cover any part of the load demand, provided there is enough water in their reservoirs. They should participate essentially in the base-load coverage with their 'run-of-river' power, but their alpine characteristics should not be forgotten. The latter means that they are able to generate more energy in the summer and less during the winter period. Hydro plants may be started within tens of seconds, and they have practically constant efficiency within the entire range of power output, which may be regulated widely.

Gas turbine plants have the advantage of low investment costs and short construction lead times, but they rank behind hydro plants with regard to rapid starting, breakdowns and operability over the entire load range owing to low efficiency. In addition, being pure peaking installations, they are less suited to longer periods of operation, have a relatively high heat rate and are also not so advantageous with respect to the environment because of noise and waste gas.

Gas turbines average approximately 2% annual load factor and run on average for approximately 1–25 h per start-up. In addition, they have proved their capability to come on-load quickly, 2–3 min to full output, to meet rapid demand variations or sudden loss of generation. Gas turbines in separate stations should have the facility for their alternators to run declutched to provide synchronous reactive compensation.

Some gas turbines are installed in separate gas turbine stations connected directly to lower voltage systems. Accordingly, they can act as auxiliaries in the new large fossil-fired stations and AGR nuclear stations. The auxiliaries in the fossil-fired stations allow a station to restart if it is shutdown and isolated from the system by a fault, and in addition their output is regarded as available to meet peak demands. By contrast, auxiliary gas turbines in AGR stations are at present regarded as being there for station security purposes only and do not provide firm capacity towards meeting the demand. If the proportion of gas turbines is increased rapidly, then their average load factor will increase. Their average operating costs will then increase rapidly not only because they burn expensive distillate oil, but also because their maintenance costs, which are a considerable part of the operating costs, increase with operation. Furthermore, the effects of demand variability (from weather or forecasting error) on gas turbine load factors are non-linear. A new gas turbine planned to have a mean load factor of 2% in a given winter might fall to 0% load factor in a mild winter with low demand – but rise to 10% in a severe winter, incurring high operating costs.

The shorter construction lead time and smaller size of separate gas turbine stations, however, do mean that it will always be convenient to include some gas turbines from time to time in new plant programmes.

Renewable energy sources – tidal, solar and wind particularly – will play a significant role in the supply side structure of future power systems. Their intermittent nature may be smoothed partly by reserve capacity of the power system, partly by finding a special place for them in the load demand curve and partly by using secondary storages. Wind and tidal sources are expected to participate in base generation, while solar sources are only suited to the intermediate zone of generation curve.

Conventional thermal plants, fired by coal, oil or natural gas, are, in sufficient numbers, perfectly capable of fulfilling the technical conditions of peaking power. But they can assume this duty only at considerable operating and economic disadvantages. In comparison to gas turbine installations, medium-load plants need higher investment (and thus have higher fixed costs), but markedly lower fuel costs. However, the desired flexibility and availability require a correspondingly large number of units independent of the load demand, on account of the rates of output variation attainable by the separate machines. In consequence, it becomes necessary to operate the various units at part load, which increases fuel cost. Furthermore, it is also not possible to control conventional units over the entire power range. For this reason they must be kept on line at part load for peaking duty. Operating conventional thermal units for peaking purposes thus results in efficiency losses and a restriction on optimising the possibilities with regard to the use of primary energy.

Despite some public controversy due to Three Mile Island (USA), Chernobyl (USSR) and Fukusima (Japan) nuclear plant disasters, of which the main types are briefly described in the Table 1.3, are playing a significant role in the supply side structure of the majority of power utilities worldwide (see Table 1.4.).

The presence of a large number of nuclear power plants influences the structure of load demand coverage (as may be seen from Table 1.5) and, in consequence, causes peak-load problems for two reasons:

1. The high capital costs of nuclear power call for cheap peak power owing to the requirement of a high utilisation factor for nuclear energy.
2. Large unit sizes with the resulting reserve problems make a corresponding increase of available peak power necessary.

The operation of base-load plants, such as nuclear power stations, large coal- or oil-fired plants, for non-basic load coverage represents the most unfavourable solution from an economic standpoint. Base-power installations have high fixed costs but particularly low operating costs. Keeping power from these plants available for covering peaks, with a correspondingly low utilisation factor, means either the addition of further base-load plants to cover this load category, and thus burdening the peaking power so obtained with the high fixed cost of these plants, or, within the limits of available reserve capacity, the operation of medium-load plants to make up for base-load generation lost, and, in consequence, the use of relatively expensive primary energy to the detriment of the peaking power required.

The unit size of thermal plants grew by a factor of 6 (150–910 MW), which was in the same proportion as the growth of the spinning reserve requirement.

Thermal plants, relying on a favourable specific heat rate and the availability of comparatively cheap primary energy from newly built refinery plants, lend themselves to operation in both the base-load (winter diagram) and peak-load (summer diagram) ranges.

But the possible changes in the structure of the primary energy base called for by the exigencies of political economy might have led to political restrictions, which in the first instance affect oil and natural gas power plants, which are better suited operationally for peaking duty.

All nuclear plants and most, but not all, of the 500 and 660 MW fossil-fired units operate at base load, i.e. to the limit of their availability. The merit order position of the various fossil-fired units depends on the fuel burnt (coal or heavy fuel oil), fuel transport and handling costs and the plant thermal efficiency, which varies with size and age of unit. Units between 100 and 375 MW operate at intermediate load factors, and smaller units, mostly 30 and 60 MW, operate at low load factors that approximate to peaking duty. In addition, some demand is being met from external supplies including international interconnections such as the cross-Channel link.

Table 1.3 Major types of nuclear reactors

Type	**LWR**	**SGHWR**	**Magnox**	**AGR**	**CANDU**	**HTR**	**FBR**
Meaning of acronym	Light water reactor	Steam generating heavy water reactor	Magnesium alloy fuel cans	Advanced gas cooled reactor	Canadian deuterium natural uranium	High temperature gas cooled reactor	Fast breeder reactor
Coolant	Water	Water	Carbon dioxide	Carbon dioxide	Heavy water	Helium	Sodium
Moderator	Water	Heavy water	Graphite	Graphite	Heavy water	Graphite	^{238}Uranium

Table 1.4 Nuclear power stations in the EU, 1986

	Germany	France	UK	Spain
Total electricity generation (GWh)	403.03	362.17	300.80	128.57
Nuclear generation (GWh)	119.30	254.14	58.78	37.46
Percentage of nuclear in total electricity production	29.60	70.20	19.50	29.10
Capacity factor	0.72	0.65	0.66	0.76
Net capacity (MW)	18.95	44.69	10.22	5.60
Gas cooled reactors	–	1.74	4.16	480
Light water reactors	18.62	41.52	–	5.12
Fast reactors	17	1.43	234	–
Advanced gas cooled reactors	309	–	5.83	–
Plant under construction (MW)	4.05	17.81	2.52	1.92
Plant planned (MW)	13.44	1.45	1.18	4.01

Each of the types discussed has its own place in the generation curve to meet the load demand optimally. The particular features of each type of generating unit are summarised in Table 1.6, consideration of which may, hopefully, enable optimal selection of units to meet the demand.

A picture of the overall performance of a particular generation system can be given by the generation duration curve, which is obtained by superimposing layers as high as the maximum capacity and as wide as the utilisation duration of each unit, arranged according to decreasing duration.

The sharing of capacity to be installed among the different types of units can be approached with reference to the generation duration curve by the well-known method of determining the power ordinates corresponding to the limit utilisation hours for which the overall annual costs (fixed plus production costs) per unit of capacity on two types of units becomes equal.

The characteristics of the different units chosen to build up the supply side of a power system affect the shape of the generation duration curve so that the calculation procedure for sharing the installed capacity has to be iterated, starting either from the existing generation set or from a supply side structure chosen on an expert basis.

The amount of the reserve margin at this stage of calculation has to be implicitly assumed as known; in fact, it must be calculated with reference to some indices that measure the risk of failure in the load coverage. Such a risk depends on the forced outage rate of different units so that the maximum height of the generation duration curve should also be determined by an iterative procedure.

The reference to the load demand curve alone, even in the chronological hourly stepwise form, takes into consideration only the constraints arising from the load side. Such an approach may be utilised for such problems as the unit's commitment or load dispatching in the planning stage of the power system expansion. A separate evaluation of the amount of reserve requirement, which is related to the economic scheduling in production and to the load supply, is essential for a comprehensive appraisal of the system operation cost, otherwise the result will not be sufficiently correct.

Table 1.5 Types of generating units available for the supply side

Plant type	Nuclear	Thermal lignite	Oil fired	Gas fired	Coal fired
Plant role	Base	Base		Intermediate	
Relative capital cost	4.4	3.3			2.8
Primary energy source	Uranium or plutonium	Lignite	Oil	Gas	Coal
Primary energy consumption (Mtol/kWh)	11	11–14	9	9	10–12
Relative fuel cost	0.3	0.3–0.45	1.4	1.3	10
Starting times ignition to synchronisation	5.5 h	5.5 h cold start 40 min hot start	75 min	75 min	75 min
ignition to full load	8.5 h	8 h cold start, 90 min hot start	90–120 min	90–120 min	90–120 min
Pick-up rate (%/min)	0.1	2	4	5	3
Controlling range (min/max)	0.8–1	0.6–1	0.25–1	0.25–1	0.3–1
Lifetime (years)	15	30	30	30	30
Environmental pollution	Waste: heat and radiation	Waste: heat and fuel gas	Waste: heat and fuel gas	Waste: heat and fuel gas	Waste: heat and fuel gas

Hydro		Renewable		Combustion	
Hydro	**Tidal**	**Wind**	**Solar**	**Industrial gas turbine**	**Aero-engine gas turbine**
Base peak	Base	Base	Intermediate	Peak	Peak
				1.0	
Water	Water	Wind	Solar radiation	Gasoline	Aero-engine fuel
				10	10.12
				1.3	1.5
				5.8 min	2 min
				4.6 min in emergency, 8–17 min normally	1 min in emergency, 5 min normally
50				30	30
Full range 50	Full range 50	Full range 20	Full range 20	0.2–1	0.2–1
				15	15
		Noise		Waste: heat and fuel gas	Waste: heat and fuel gas

Table 1.6 Supply side structure in proportion to declared net capability for different power systems

Type of generating capacity	Germany	France	England and Wales	Spain	Russia		Ukraine
					Central	Northwest	
Nuclear	0.178	0.422	0.097	0.147	0.213	0.235	0.185
Conventional thermal	0.749	0.333	0.861*	0.491	0.712	0.627	0.726
Hydro	0.073	0.245	0.042**	0.362	0.075	0.138	0.089

*Includes 0.056 installed in gas turbines.
**Includes 0.04 installed in pumped hydro.

The optimum composition of the generating system should derive from consideration of the interaction between the structure and the performance characteristics of the supply side, on the one hand, and the behavioural characteristics of the demand side of the power system, on the other.

Typically, between 40% and 60% of a system's base load is supplied by large coal and nuclear units of the highest efficiency burning fuel of the lowest cost. Such base-load units are operated continuously for most of the year. The broad daily peak in demand, representing another 30–40% of the load, is met by 'cycling' or 'intermediate' generating equipment, usually the system's less-modern and less-efficient fossil-fuel (coal, oil or gas) units, hydroelectric power units where they are available and gas turbine units where they are needed. Although the electricity generated by cycling plants costs more than base-load electricity, such plants are the most economical way of generating electricity for part of the time every day, adding up to perhaps 1500–4000 h per year. Sharp peak demands are met by still older fossil-fuel units and by hydroelectric power-, gas- or oil-fired turbines and diesel generators. Such units operate for a few hundred hours per year to 1500 h.

The structure of the supply side of some European and Russian utilities is given in Table 1.7, while the individual contributions to load demand coverage are shown in Figure 1.4.

This traditional three-level composition of generating plants has become increasingly less attractive as sharply rising fuel costs penalise the less-efficient units. Moreover coal-fired units now require expensive but essential pollution control equipment that represents an economic disincentive to cycling operation. So, new non-traditional compositions are needed.

There are no unique criteria, if technical characteristics are excluded, for reaching an optimal supply side plant structure. The economically optimal solution will vary depending on geographical conditions, environmental considerations, primary energy availability, financing problems, composition of supply side already available and so forth.

Table 1.7 Power system general problems and conventional ways of solving them

Problem	Duration of the fluctuation	Conventional way of solving problem
Stability improvement	0.02–0.2 s	Using PSS and shunt resistor brakes
Countermeasure against blackout	0.12–0.2 s	Cutting off the damaged part of the system; using shunt resistor brakes
Voltage stabilisation		SVC
Frequency control	0.5–120 s	Boilers governor, steam reserve in fossil plant
Spinning reserve	30–300 s	Part loaded plants
Peak lopping	180–10 000 s	Gas turbines, low merit fossil plant, intersystem links
Load leveling	Daily 4–12 h, weekly 40–60 h, seasonal 3 months	Mid merit fossil plant, spare plant, scheduled maintenance
Output smoothing	Intermittent	Power system spinning reserve, stand-alone diesel generator

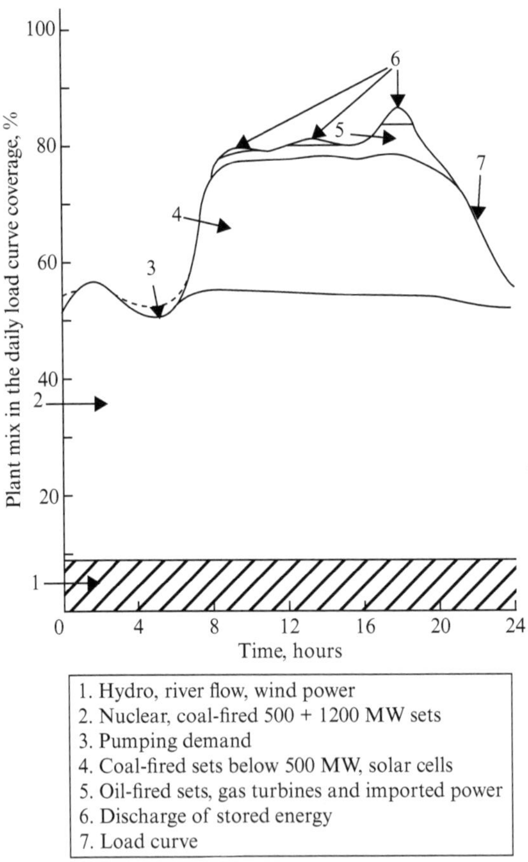

Figure 1.4 Plant contribution for typical winter-day load demand coverage

1.3 Generation expansion planning

Generally speaking, any power system comprises three parts: namely supply side, distribution system and demand side. It is a complex combination of installed nuclear, thermal, hydro and other renewable electricity generators that are interconnected with commercial, industrial and domestic loads by transmission lines of different length and voltage. An outline of a power utility structure is shown in Figure 1.5.

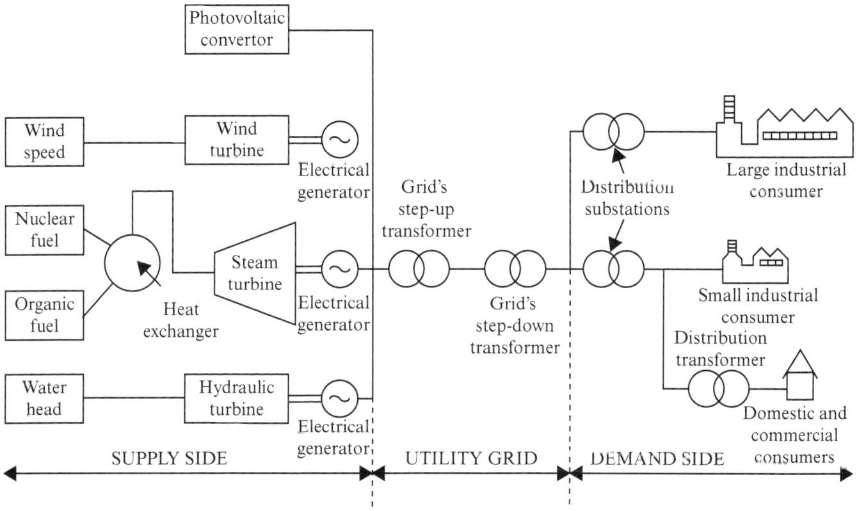

Figure 1.5 Outline of a power utility structure

The structure of any power system is not solid. There are at least two reasons for this. First, load has an intermittent nature and is continuously growing. The other is concerned with the expansion of existing plants in order to face the growing demand. In addition, old power plants need to be retired and new ones constructed. The economic value of the participation of a new unit in a given generating system has to be considered in relation to the variations imposed on the energy production of the existing units and to the reserve requirements.

However, of more relevance to the problem of planning to meet future load demand is not the rated power of new plant required but the choice of plant type. In the long-term planning of a generating system, it is important to find a correct policy for the introduction of new units. New plant must be justified on its overall economic merit and not just on a requirement to cover a certain part of a load demand curve.

The capacity of new generating plant required to meet future demand depends on a number of factors. The construction time for the first unit in a new main generating station of established design is six years, so that if such a station were ordered in 2010 its first unit would be commissioned in time to contribute to meeting the load demand in 2017, the seventh winter ahead. The first step in the planning process is therefore to predict the typical parameters of load demand for at least the seventh year ahead.

With regard to these characteristics, there is a constant design problem in the power system: to determine the optimal generation expansion strategies – optimal type, capacity and site for each new introduced power unit – to meet forecast load demands with a certain reliability over a planning horizon of approximately seven years or more. The general statement of the optimisation problem is well known: to minimise the present value of the total cost that consists of capital, operation and risk components.

To find the optimal solution of a power system expansion-planning problem means to choose an expansion alternative that minimises the power system cost function including capital costs and relevant fuel costs for the generation equipment as well as costs of losses in the power system components and in transmission lines. Obviously the selected optimal alternative must satisfy a multitude of criteria – technical, financial, environmental and social.

Mathematically this optimisation problem may be written as

$$\min f(x) = CX$$

$$\text{subject to } AX < b$$

where $f(x)$ = the power system cost function mentioned above
A, b = coefficients from the set of technical and environmental constraints
X = vector of variables to be found as an optimal solution.

The costs associated with each alternative plant are calculated over its lifetime: these are capital and other fixed costs, fuel cost, other direct operating costs and the effect on the cost of operating other plant on the system. This calculation requires simulation of the operation of an expanding system over a number of years ahead and involves the use of optimising techniques.

As far as the methods of calculation are concerned, there are two main approaches. The first aims to solve the optimisation problems using such calculation methods as linear programming and, in particular, integer linear programming. A dynamic programming method with sub-optimisation may also be suggested. The random characteristics of the load demand have to be considered. Separate reserve evaluation, reflecting the random characteristics of the generation system, is associated with this approach.

The second kind of approach is that of Monte Carlo simulation of system operation, which 'simultaneously' takes directly into account the demand and generation of random characteristics including those related to the energy availability of the hydro subsystems. Optimisation is then obtained by parametric analysis.

Finally, the application of standard discounted cash-flow techniques to the cash flows for each project allows an economic comparison to be made.

1.4 Meeting the load

Since the structure of the generating side of the power system is designed to face the maximum foreseen demand, which in fact might never occur – normal everyday demand is much less than the maximum – the dispatch problem can be stated: to find the optimal load for each generation unit to face the demand forecast for the next day and to adjust this solution to the real current demand.

The elements that form the structure of the supply side have to meet variations in the demand side load. As is clear from the daily load curve in Figure 1.6, the main problems are as follows:

- unloading generating plant at the beginning of the night trough;
- loading plant sufficiently rapidly in the morning rise period;
- meeting demand variations during the day, and particularly the evening peak; and
- meeting a sudden loss of generation or an unexpected increase in demand.

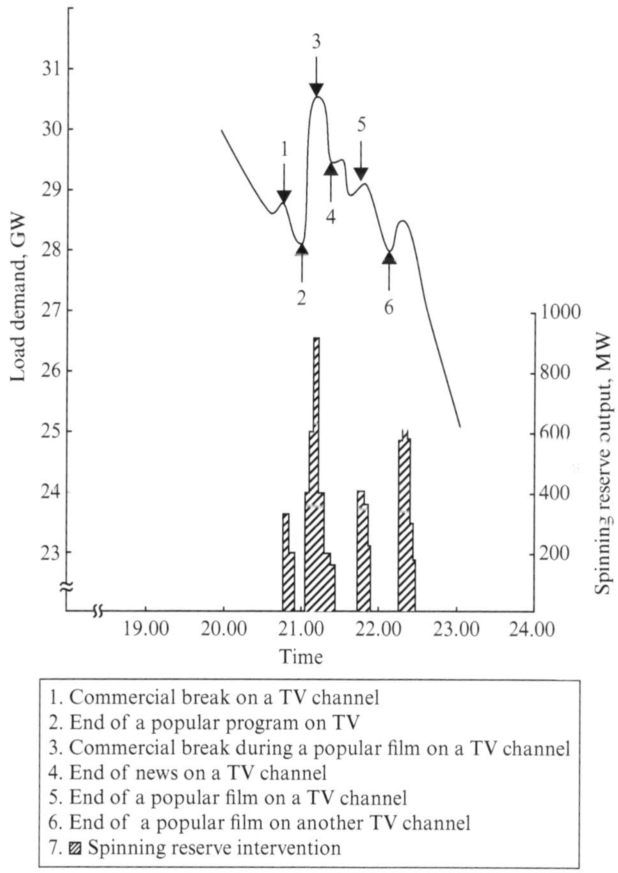

Figure 1.6 Typical winter evening load demand curve when a popular TV program may cause a demand increase late in the evening

Such problems as stability of voltage and frequency control are also in one way or another connected to variations in the load demand.

All of these problems are more or less successfully solved in power utilities by using a wide range of special devices and methods, which are collected together in Table 1.7.

In following the continuously variable load curve, one must take into account the operating and risk costs related to change of output, to spinning and ready standby reserve and to control.

One may conclude that the need to meet variable loads, especially peak ones, causes the vast majority of power system problems.

The term 'peak load' actually describes different states of system load:

- on the one hand, the absolute maximum load in the course of a day, month or year and
- on the other, the short-time relative load peaks in the course of the daily load diagram.

The peak duty in the generation system derives not only from the need to cover peaks of the load diagram, but there is also an important element to cover the need to cope for short durations with the troughs in the diagram of available generation owing to their forced outages.

Common to both states is the low utilisation of plant capacity required. The technical and operating demands on the coverage of these different types of peak load vary widely.

The requirements for peaking operation are the following:

- wide output variation rates;
- low starting losses;
- short run-up times with low stress on the plant – also high availability if frequent starts and stops are involved;
- rapid operability; and
- automation possibilities and simple handling.

Comparing these requirements with the information given in Section 1.2, it is not difficult to conclude that most of them are not met by a significant part of the generating equipment. A new generation of power equipment is therefore needed and the most promising seems to be new and different types of energy storage devices.

Chapter 2

Energy storage as a structural unit of a power system

2.1 General considerations

The diversity of applications of electricity and particularly the fact that some of its uses, such as lighting and space heating, are subject to substantial seasonal variation makes the economic ideal of supply for constant consumption throughout the year unrealistic. Generation itself cannot, in any case, be constant because of fluctuations in hydroelectric generation and intermittency of renewable sources.

There should therefore be an intermediate unit between producer and customer that can coordinate them. This unit has to provide the following two possibilities:

- for producers, to transfer generation or production capacity from off-peak to peak load hours to supplement the development of specific peak-production means;
- for distributors or customers, to encourage customers to shift peak hour consumption requirements to off-peak times. (Incidentally, the customer can also alter his/her habits.)

This intermediate unit therefore has to be able to separate partly or completely the processes of energy generation and consumption in the power system. We will call this unit secondary energy storage

> *Secondary energy storage in a power system is any installation or method, usually subject to independent control, with the help of which it is possible to store energy, generated in the power system, keep it stored and use it in the power system when necessary.*

Secondary energy storage (ES) is an installation specially designed to accept energy generated by the power system, to convert it into a form suitable for storage, to keep it for a certain time and to return as much of the energy as possible back to the power system, converting it into a form required by the consumer.

Secondary energy storage method is a load management strategy that helps utilities to reduce peak load and to shift some of the required energy to base load power plants.

According to the above definition, energy storage may be used in a power system in three different regimes:

- charge;
- store; and
- discharge.

In each of these three regimes a balance between power and energy in the power system has to be maintained so the energy storage has to have the appropriately rated power and energy capacity.

The duration of each regime, its time of reverse (switching time) and storage efficiency are subject to power system requirements.

To fulfil the requirements given in this definition any complete energy storage unit must contain three parts:

- power transformation system (PTS);
- central store (CS); and
- charge–discharge control system (CDCS).

The PTS has to couple the power system and the CS. It also acts as a power conditioning system and controls energy exchange between CS and the power system. Basically there are three different types of PTS, namely thermal, electromechanical and electrical (see Table 2.1).

There are two possible ways to couple energy storage to power system: parallel connection and series connection (see Figure 2.1). In the case of series connection, energy storage also has to act as a transmission line, so its rated power has to satisfy the system requirements for these lines: all the generated energy passes through its PTS. For parallel connected energy storage, the power exchange between the CS and the power system passes through the PTS, so the rated power of the latter element has to satisfy the power system requirement for the energy storage power capacity P_s. For the same reason, the variable characteristic of the PTS has to satisfy the power system's requirement for the time of reverse t_{rev}.

The CS comprises two parts: storage medium and medium container. There are the following types of CS (see Table 2.1):

- thermal, using sensible or latent heat of the relevant storage medium;
- mechanical, using gravitational, kinetic or elastic forms of energy;
- chemical, using chemically bound energy of the storage medium; and
- electrical, using electromagnetic or electrostatic energy of the relevant storage media.

The CS is only a repository of energy from which energy can be extracted at any power rate within the PTS installed capacity margins until it is discharged. The CS should be capable of charge (discharge) up to a predetermined power level throughout the entire charge (discharge) period t_W. In practice, this means that a certain part of the stored energy must be kept in the CS to ensure the ability of the PTS to work at a required power level. It should be mentioned here that no single type of CS allows the full discharge of stored energy without damage to the whole installation.

The CDCS controls charge and discharge power levels in accordance with the requirements of the power system's regime. It is an essential part of any storage

Table 2.1 Types of secondary energy storage

Storage type	CS		PTS	
	Storage medium	**Storage vessel**	**PTS**	**Control parameters**
Mechanical				
Pumped hydro	Water	Upper and lower basins	Motor generator driven pump turbine	Water valves, motor generator's excitation current
Flywheel	Rotating mass	Rotating mass	Motor generator	Excitation current of motor generator
Compressed air	Air	Artificial or natural pressurised air containers	Motor generator driven compressor turbine	Air valves, motor generator's excitation current
Sensible or latent heat	Water, gravel, rocks, etc.	Different types of artificial or natural containments	Conventional thermal power plant	Steam and water valves, generator's excitation current
Chemical				
Synthetic fuels	Methane, methanol, ethanol, hydrogen, etc.	Fuel containers	Any conventional thermal power plant	Thermal plant conventional control means
Secondary batteries	Couple of electrodes–electrolyte system	Battery casing	Thyristor inverter/rectifier	Inverter/rectifier's firing angles
Fuel cells	Synthetic fuels such as hydrogen	Fuel cell's casing	Electrolyser plus inverter/rectifier	Inverter/rectifier's firing angles
Electrical				
Super-conductive magnetic storage	Electromagnetic field	Super-conductive coil	Thyristor inverter/rectifier	Inverter/rectifier's firing angle
Capacitor	Electrostatic field	Capacitor	Thyristor inverter/rectifier	Inverter/rectifier's firing angle

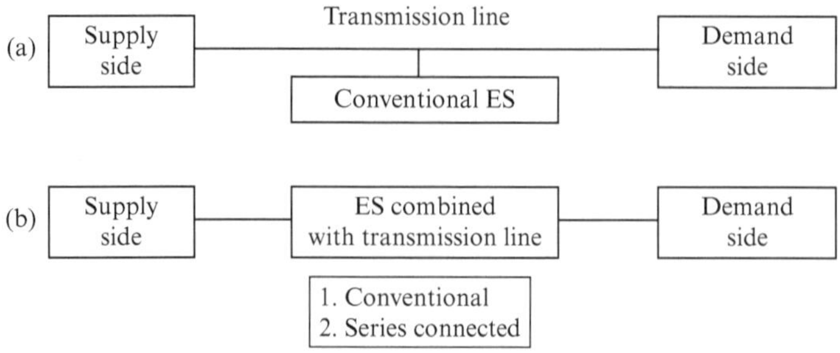

Figure 2.1 Schematic diagram of the position of an energy storage device within a power system. (a) Conventional (b) Series connected

device and usually comprises a number of sensors placed in certain nodes of the power system, in the PTS and in the CS. The information from these sensors has to be collected and used in a computer-based controller which, using relevant software, produces commands for power flow management in the PTS.

It is clear from Table 2.1 that different types of storage equipment use different physical principles, for which reason direct comparison of storage systems tends to be very complex. Therefore, it would be reasonable to select a number of energy storage characteristics common to any type of equipment and make a comparison using these. The following general features are the key parameters for comparison when discussing storage systems:

- energy density per mass and volume;
- cycle efficiency;
- permissible number of charge–discharge cycles;
- lifetime;
- time of reverse and response time level;
- optimal power output;
- optimal stored energy; and
- siting requirements.

Before installing any device in a power system a planning engineer should decide in what way a utility is going to use it – the function this device will perform. We will define the power system requirements for energy storage as a margin for its rated power P_s and energy capacity E_s, its efficiency Q_s, switching time t_{rev} and duration t_w of its regimes. It is clear that these margins – power system's requirements – are subject to the function performed by energy storage in a power system.

Energy storage could be deployed for one or more of the following reasons:

- to secure power system generating capacities optimal remix;
- to improve the efficiency of a power system operation;
- to reduce primary fuel use by energy conservation;
- no alternative energy source available; and
- to provide security of energy supply.

There are two different types of energy systems – hybrid and combined systems. We shall define a hybrid energy system as a system with one kind of energy output and two or more energy sources as input. The combined system, in contrast, has one primary source as input and two or more different kinds of energy output.

The hybrid system has been used in the transport sector. The combined system has been widely used in combined heat and power generation, where a power station utilises the waste heat from electricity production as district heating.

Energy storage usage is possible in both these systems.

2.2 Energy and power balance in a storage unit

As it has been mentioned earlier, any storage facility designed for use in a power system comprises three parts (see Figure 2.2):

- central store (CS);
- power transformation system (PTS); and
- charge–discharge control system (CDCS).

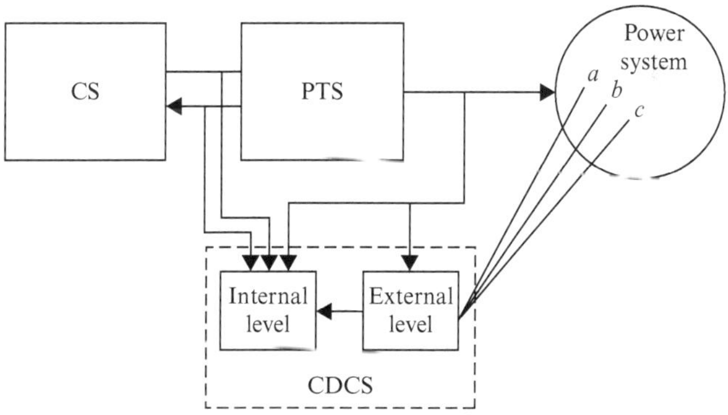

Figure 2.2 Schematic structure of energy storage a, b, c = selected nodes

The CS can be completely defined 'from a power systems point of view' by its energy capacity E_s – the energy storable in the storage vessel. This obviously has a different nature according to the type of CS – mechanical, thermal, chemical or electrical. It should be mentioned that a certain part of E_s has to be kept in the CS to ensure the possibility of it being charged or discharged at a designed power capacity P_s, which is the second main characteristic of the storage facility.

The PTS can be defined by its rated charge or restoring capacity P_c and discharge or generating capacity P_d. Power capacity P_s equals the maximum of P_c and P_d:

$$P_s = \max \{P_c, P_d\}$$

The CDCS is a specially designed computer-based controller with the corresponding software.

Since energy storage is a part of the power system, it has to work in all its normal and emergency regimes. In any of these regimes, the power and energy balance in the node, where storage is coupled to the power system, has to be maintained:

$$N_{gen} - L_l + P_s = 0$$
$$Q_{gen} + Q_l + Q_s = 0$$

where N_{gen}, Q_{gen} = active and reactive power generated by the supply side of the power system
L_l, Q_l = active and reactive power consumed by demand side of the power system
P_s, Q_s = active and reactive power from store

Note that the power generated by or discharged from the store is positive whilst the power consumed by or charged into the store is negative (see Figure 2.3).

The energy balance for energy storage (see Figure 2.4) reflects the fact that storage is only a repository of energy and not an ideal one. There are certain losses in the storage δE_s, and it is possible, taking them into account, to write the energy balance equation

$$E_{gen} - \delta E_s - E_l = 0$$

where E_{gen}, E_l = energy generated and consumed by the power system, respectively.

The energy losses δE_s comprise the losses during the charge, store and discharge regimes shown in Figure 2.3:

$$\delta E = \delta E_c + \delta E_s + \delta E_d$$

or a difference

$$\delta E = E_c - E_d$$

where E_c, E_d = energy charged and discharged, respectively.

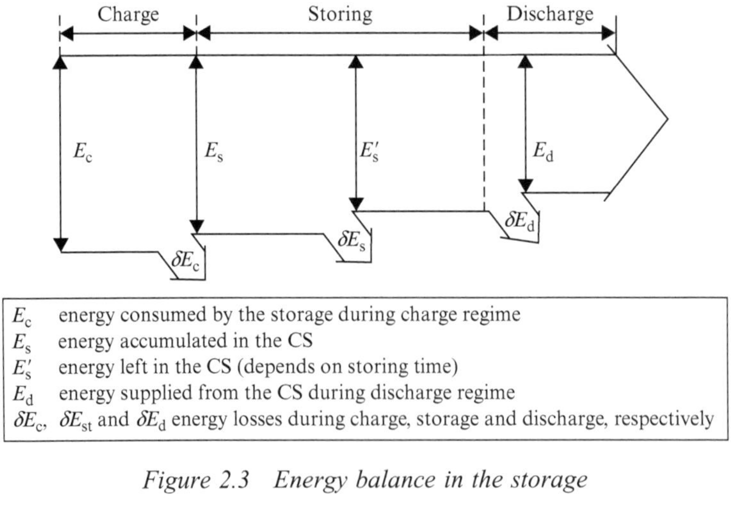

Figure 2.3 Energy balance in the storage

We will define charge efficiency ξ_c as

$$\xi_c = \frac{E_s}{E_c}$$

storage efficiency $\xi_s(t)$, which depends on the duration of this regime, as

$$\xi_s(t) = \frac{E_s'}{E_s}$$

and discharge efficiency ξ_d as

$$\xi_d = \frac{E_s'}{E_d}$$

Obviously, the overall cycle efficiency ξ_s, which is the ratio

$$\xi_s = \frac{E_l}{E_{gen}} = \frac{E_d}{E_c}$$

can be defined as a result of multiplication

$$\xi_s = \xi_c \xi_s(t)\, \xi_d$$

Energy losses δE_s can be written as follows:

$$\frac{\delta E_s = E_c - E_d = E_s}{\xi_c / E_d} = \frac{E_s(1 - \xi_c E_d / E_s)}{\xi_c} = \frac{E_s(1 - \xi_c \xi_s(t) \xi_d)}{\xi_c}$$

or

$$\frac{\delta E_s = E_s(1 - \xi_s)}{\xi_c}$$

2.3 Mathematical model of storage

Mathematical modelling of energy storage requires the formulation of a model for each of its three elements – CS, PTS and CDCS.

Since the CS is an energy repository, its only function is to accumulate energy in the desired quantities and release it at a predetermined speed. The stored energy E_s is a function of the following types of parameters:

- construction parameters CP_i, which are constant for the given energy storage device;
- variable parameters VP_i, which depend on the current regime of the storage; and
- current time t.

The function may be given as follows:

$$E_s = f_{CS}(CP_i, VP_i, t)$$

According to the definition, power is a first derivative of energy with respect to time. Therefore, the power flow from the CS equation may be given by

$$\frac{P_{\mathrm{CS}} = dE_s}{dt} = \frac{df_{\mathrm{CS}}(\mathrm{CP}_i, \mathrm{VP}_i, t)}{dt}$$

The main function of the PTS is to regulate the power flow from the energy storage in accordance with the reference power system requirements. The power flow through the PTS depends on the same variable parameters VP_i. In addition to this, the power flow depends on the special regulation parameters (RP_i), the constant PTS's parameters CPP_i and the power system regime parameters PWSRP_i in the reference node. The mathematical model of the PTS equation may therefore be given by

$$P_{\mathrm{PTS}} = f_{\mathrm{PTS}}(CPP_i, RP_i, PWSRP_i, VP_i)$$

The CDCS has to measure the regime parameters in the given nodes of the reference power system and in the energy storage, calculate the desired power flow P_{des} from energy storage, then on this basis calculate the value of the special regulation parameters and send them to the PTS. It should be mentioned that one of the regulation parameters controls the charge, store or discharge mode of the energy storage.

The mathematical model of the CDCS equations may be given by

$$\begin{aligned} P_{\mathrm{des}} &= f_{\mathrm{CDCS}}(PWSRP_i) \\ RP_i &= F_{\mathrm{CDCS}}(P_{\mathrm{des}}, VP_i) \end{aligned}$$

The energy storage is not a source of energy, so an energy balance equation is required. In this equation storage efficiency ξ_{s} has to be taken into account, so that it may be given by

$$\frac{E_{\mathrm{c}} - E_{\mathrm{d}} - E_{\mathrm{s}}(1 - \xi_{\mathrm{s}})}{\xi_{\mathrm{c}} = 0}$$

where E_{c}, E_{d} = energy charged and discharged from the CS.

The full mathematical model of energy storage may therefore be given by the following system of equations:

$$\begin{aligned} P_{\mathrm{CS}} &= \frac{df_{\mathrm{CS}}(CP_i, VP_i, t)}{dt} \\ P_{\mathrm{PTS}} &= f_{\mathrm{PTS}}(CPP_i, RP_i, PWSRP_i, VP_i) \\ P_{\mathrm{des}} &= f_{\mathrm{CDCS}}(PWSRP_i) \\ RP_i &= F_{\mathrm{CDCS}}(P_{\mathrm{des}}, VP_i) \\ P_{\mathrm{CS}} &= P_{\mathrm{PTS}} = P_{\mathrm{des}} \end{aligned}$$

It should be mentioned that the particular type of the functions $f_{\mathrm{cs}}, f_{\mathrm{PTS}}, f_{\mathrm{CDCS}}$ and F_{CDCS} depend on the type of the CS, the PTS and the number of desired duties that the energy storage has to perform in the power system.

Use of the full mathematical model can only be justified when a particular type of energy storage is involved and optimisation of its parameters or optimal control is needed.

If a system requirement for energy storage is involved, it is possible to simplify the full model, leaving only the following equation:

$$P_{\text{des}} = f_{\text{cdcs}}(PWSRP_i)$$

It is assumed here that $P_{\text{des}} = P_{\text{CS}} = P_{\text{PTS}} = P_{\text{s}}$ and that the only information needed about RP_i is information about the storage mode.

The operating characteristics of active and reactive power controls by energy storage are considered, as the transfer characteristics from the desired values δP_{des}, δQ_{des} to be the actually controlled values δP_{s}, δQ_{s}. They are modelled as independent first-order time lags.

$$\delta P_{\text{des}} = \frac{\delta P_{\text{s}}}{(1 + T_{\text{s}}P)}$$

$$\delta Q_{\text{des}} = \frac{\delta Q_{\text{s}}}{(1 + T_{\text{s}}P)}$$

The storage time constant T_{s} has to be determined for each particular storage technique.

2.4 Econometric model of storage

The capital cost C_{s} of energy storage is the sum of two parts. One is related to the storable energy; the other depends on the peak power that the storage must deliver and is controlled by the CDCS according to demand requirements. In other words, the capital cost depends on the cost of CS, PTS and CDCS, partly proportional to installed power capacity and in part to storable energy capacity:

$$C_s = C_{\text{CS}} + C_{\text{PTS}} + C_{\text{CDCS}}$$

It is convenient to use the specific cost per unit of storable energy capacity and installed power capacity, so that usually the capital costs for the storage components are given by

$$C_{\text{CS}} = C_{\text{e}}^{*} E_{\text{s}}$$

$$C_{\text{PTS}} + C_{\text{CDCS}} = C_{\text{p}}^{*} P_{\text{s}}$$

where C_{e}^{*} = specific cost for the CS, money/kWh
C_{p}^{*} = joint specific cost for PTS and CDCS, money/kW

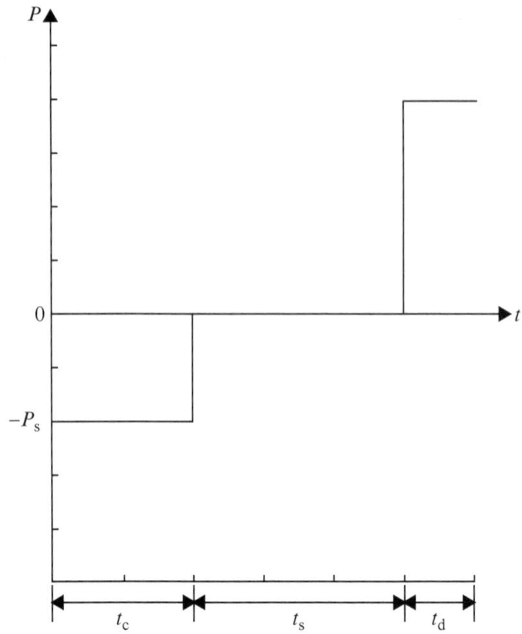

Figure 2.4 Energy storage regimes t_c, t_s and t_d: charge, storage and discharge durations, respectively

Hence the capital cost of energy storage is a function of the two main storage characteristics and may be written as follows:

$$C_s = C_e^* E_s + C_p^* P_s$$

If reference is made to the specific cost per unit of generating capacity, which is widely used in a power system design analysis, this cost may be given as follows:

$$\frac{C_s}{P_d} = \frac{C_e^* E_s}{P_d} + \frac{C_p^* P_s}{P_d}$$

or

$$C_s^* = C_e^* t_d + C_p^* K_u$$

where K_u = restoring ratio
t_d = rated discharge time

The annual cost of a storage facility, Z_s, when operational cost comprises capital repayments, interest charges, operating and maintenance costs plus the cost of energy losses, may be given as follows:

$$Z_s = RC_s + n_c C_u \delta E_s$$

where R represents all payments that depend on capital cost – capital repayments, interest charges, operational and maintenance cost (in other words the constant part of the annual cost), n_c is the number of charge–discharge cycles during a year and C_u is the specific cost of energy used for charging the storage. The cost of annual energy losses is a variable part of the annual cost.

The econometric model of an energy storage facility is usually given in the form of the annual cost of the facility:

$$\frac{Z_c = R(C_e E_s + C_p P_s) + n_c C_u E_s (1 - \xi_s)}{\xi_c}$$

which allows us to optimise its key parameters – installed capacity and stored energy – or to use this model as a part of the power system econometric model.

Energy storage is part of a power system and therefore P_s and E_s have to satisfy the special system requirements. To define these requirements, it is necessary to formulate a power system mathematical model in which the energy storage model will be included.

Chapter 3
Storage applications

3.1 General considerations

A typical electricity bulk supply power system consists of central generating stations (supply side) connected to a transmission system. This bulk supply system is joined to the distribution system that comprises a subtransmission system of primarily distribution feeders and secondary circuits.

An energy storage unit can be connected to the transmission, subtransmission or distribution system in a manner similar to customer-owned conventional or renewable generation facilities such as gas or wind turbines. These dispersed sources are able to change the character of a typical electricity power system completely.

With any addition of dispersed generation facilities, whether they are customer-owned or utility-owned, the impact upon planning, control, protection and operation of the traditional distribution system and the bulk power transmission system has to be closely investigated.

In an electricity power system based on thermal, nuclear, hydro and renewable generation, storage will find a wide field of application and may perform various duties, which must be taken into consideration in order to gain the largest possible advantage in the supply side optimisation.

3.2 Static duties of storage plant

A first examination will be given here of the duties that storage methods could perform under normal system conditions – so-called static duties.

Most, if not all, power utilities worldwide have to meet increasing demand from year to year. This means that additional plant must be installed to meet the peak demand on the system.

It is clear that an installation capable of taking electricity from the grid at night and returning it during peak periods will reduce the need for generation capacity in the system. Hence the capital cost of a storage unit may be compensated by savings on the conventional power stations, which will be superseded.

Contrary to popular belief, the variation of demand on electricity utilities does not constitute a need for storage, but provides an opportunity for storage methods to compete with mid-merit and peaking generating sources. The preferred plant is that which provides the minimal balance between the capital charges and discounted operating cost savings, which the plant can make when operating in its merit position over its life. Operating cost savings are very sensitive to the expected fuel cost of the candidate plant, which determines both the generating cost margin and the number of hours in each year over which savings will be made.

The variation of load through the day stimulates a demand for storage especially when the increase in installed capacity of large coal or nuclear plants, designed to operate at maximum efficiency on their rated power output, exceeds base-load demand, and when an increase in utilisation of intermittent energy sources (such as solar, wind or tidal energy) exceeds the utilities' reserve capacities.

Consider the typical weekly load curve of a utility with and without energy storage, as shown in Figure 3.1. As illustrated by the upper curve, the intermediate and peaking power involves extensive generating capacity. The load variation shown here is typical for any European utility, but it applies to most other countries where cheap off-peak electricity rates exist. In countries where this is not the case, the daily variation tends to be even larger unless these are countries without a restricted consumption and multi-shift industry. In any case it appears to be the fact internationally that installed capacity is about double the yearly average load so the electric power industry runs at considerably low capacity factors (in the US as low as 40% – according to the Electricity Advisory Committee (EAC), a panel of experts that advises the US government on industry issues). Increased levels of

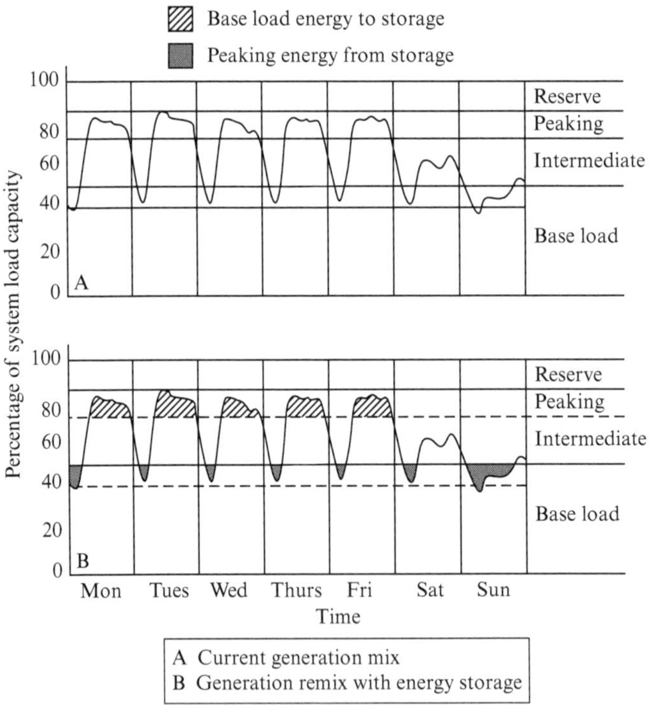

Figure 3.1 Weekly load curve for a typical power utility

renewable energy on the grid will lead to even lower capacity factors for conventional generation sources.

If large-scale energy storage were available, as illustrated by the lower curve of Figure 3.1, then the relatively efficient and economical base power plants could be used to charge the storage units during off-peak demand (lower shaded areas in Figure 3.1).

Discharge of the stored energy (upper shaded areas) during periods of peak load demand would then reduce or replace fuel-burning peaking plant capacity, thus conserving (mostly gas or oil-based) fuel resources. Use of energy storage to generate peak power in this manner is termed 'peak saving'. The higher base-load level may replace part of the intermediate generation, thus performing load leveling and enabling the more extensive use of storage to eliminate most or all conventional intermediate cycling equipment. Assuming that new base-load plants use non-oil-based fuel, there are further savings of both cost and oil resources.

The most typical longer term duty is the storage of energy generated at low incremental cost during low-load hours, and the return of this energy to the load during high-load hours, thus replacing the energy generated at high incremental cost. This duty, which we may call 'energy transfer duty', is reasonable from the economic point of view only if the ratio between the incremental cost of energy during the storage charge period and the incremental cost of replaceable energy during the storage discharge period is lower than the turnaround efficiency of the storage unit:

$$\xi_s > \frac{C_{night}}{C_{day}}$$

where ξ_s = storage turnaround efficiency
C_{night}, C_{day} = night-time and daytime marginal generation cost, respectively

The differences between the above-mentioned ratio and the efficiency should not be too small because the energy production advantage must be sufficient to compensate for the difference between capital cost for storage and additional base-load capacity, on the one hand, and the capital cost for replaced peak and intermediate capacity, on the other. In any case, a reduction in installed capacity is achieved.

Smoothing of the daily load curve also leads to reduced stresses of load following operation by the steam plant, and consequently reduces maintenance costs. The exact value of this reduction depends on a number of factors, but it is usually smaller than the fuel cost saving.

The other static duty is that of making generating power available when it is needed for matching load demand. This duty is common to all generating methods in the system but if storage methods are involved the main parameter – energy capacity – has to be taken into account.

In order to cover the period of power demand, the storage systems have to have sufficient storage capacity, and since the amount and duration of load demand are typically stochastic, unconstrained storage capacity would theoretically be necessary to avoid the possibility of being unable to cover the load owing to exhaustion of the stored energy. The capital cost of a storage system depends on its energy capacity, and therefore the choice of storage parameters should be a compromise between cost and the risk that they may not be able to cover the demand. In planning a power system it is necessary to evaluate this compromise quantitatively.

3.3 Storage at the user's level

With the help of storage methods it becomes possible to use less valuable and more economic primary sources of energy in the residential, commercial and industrial sectors. These sectors represent approximately 40% of an average EC country's oil consumption and nearly 80% of its natural gas consumption, primarily to provide heat for water, buildings and industrial processes. Shifting much of this demand to coal by burning it at the site of energy consumption would be impractical because of fuel handling and pollution problems.

The burning of oil and gas to provide heat can be reduced by shifting domestic heating loads to coal and uranium in the form of electricity, and also by turning to renewable energy. When such a strategy is effectively implemented, the importance of energy storage will increase and in some cases it will be essential.

The storage of thermal energy close to its end consumer represents another way to employ storage methods to shift energy consumption. According to this concept heat or 'cold' is produced by off-peak energy on the consumer's premises and stored for consumption during peak load periods. Since this strategy shifts the storage investment from the supply side to the consumer, utilities can and must provide financial incentives by offering tariff rates that reflect the lower costs of off-peak generation.

Through appropriate tariffs a number of US utilities have encouraged their customers to employ water heaters controlled by timers or by electrical signals from the utility as an energy storage method. This 'load management' strategy has helped utilities to reduce peak load and to shift some of the required energy to base-load power plants. The same practice has also been introduced in several European countries in recent years. In some parts of Germany, for example, storage heaters account for nearly a quarter of the total demand for electricity in winter, so that the daily load curve of a utility is nearly flat. The effect would be much greater if space heating were provided from coal- and nuclear-based power plants through the storage of heat produced by off-peak energy.

The dispatch centre in Hamburg uses remote control by means of wave trains at 283 Hz from a special control room equipped with computers for programming the emission and controlling the production of the wave trains. At Valenciennes, in northern France, remote control of storage space-heaters is coordinated by a weather-observation system, so that charging is switched on as late as possible while heating requirements are met. During the coldest days, charging takes place during the daytime for a maximum of 4 h, thus satisfying minimal consumer demand. This solution seems simpler, but it is less effective than the one adopted in

Hamburg. Further progress lies essentially in improved timing of the switching on of equipment.

The reduction of peak load demand, the shift of demand to off-peak base-load plants and the optimisation of energy losses in transmission lines may be maximised if utilities retain a measure of control over the charging of consumer-owned storage systems. Control methods currently under investigation include start and stop signals transmitted by radio, signals sent over the wires of the utility's own network and signals sent over telephone lines. Such load management does not normally inconvenience the customer, and it can reduce his/her electricity bill.

3.4 Storage and transport

The development of the transport sector has enlarged the demand for alternative fuel and energy storage. To be competitive, a good storage system must be reasonably safe, be easy to handle, operate and recharge, and have a sufficiently high energy density.

Most of the efforts to produce an alternative to petrol-driven vehicles with combustion engines have concentrated on developing an electric battery with energy density better than that of the lead-acid battery.

The high density of energy storage provided by petrol makes replacement vehicles with combustion engines quite a formidable task. An automobile tank with a volume of 3 ft^3 can store 3×10^6 Btu in the form of chemical energy, enough to give the average car a range of between 250 and 400 miles. Lead-acid batteries of this volume can store the equivalent of 20–500 Btu. It should be mentioned that approximately 40% of the energy stored in a battery is available at the driving wheels compared with approximately 10% of the energy stored in fuel, so electricity represents a higher efficiency of energy use than fuel. Nevertheless, a car driven by lead-acid batteries, with a maximum acceptable weight of approximately 10 ton, has a typical range of 25–50 miles. Although advanced batteries will have higher energy density, and consequently greater range for a given battery weight, they will be more expensive, and therefore the range limitations of EVs may well be set not by battery weight but by battery cost. This limited range, together with the high cost of chemical batteries, presents a serious barrier to the wide application of battery driven EVs.

The cost of an EV without a battery is no higher than that of a conventional vehicle of similar size. The battery cost makes the difference: the cost of a petrol tank is insignificant, but a 1-ton lead-acid battery, providing a range of 25–50 miles on a charge of 30 kWh, will add an additional 10% to the cost of the average vehicle.

The capital investment, operating costs and logistic complexities of a network of service stations in that an empty battery could be quickly exchanged for a fully charged one would almost certainly be great but on the other hand this new strategy would provide an enormous number of new jobs. The higher efficiency and continuing escalation of automobile fuel prices give EVs the potential to demonstrate that they are cheaper to operate for similar tasks.

If storage systems based on batteries could be improved sufficiently to make EVs attractive alternative to conventional cars, each million EVs would save 146 million barrels of oil per year.

Assuming that the vehicles travelled an average of 10 000 miles per year, the impact of this displacement on electricity consumption may be calculated: 1 million vehicles $\times$ 10 000 miles/yr vehicle $\times$ 5 kWh/mile $= 5 \times 10^{10}$ kWh.

Although this is not a large factor, the impact on certain utilities could be appreciable, particularly if it is necessary to charge batteries during hours of peak demand and therefore to reinforce the distribution network.

Once EVs have become widespread, power utilities will require most of the battery charging to be done in off-peak hours, when coal and nuclear base-load energy is available. In addition to setting preferential rates for off-peak charging, utilities could exercise direct control over equipment for overnight charging, and thereby integrate the charging load into the total demand of the power system, with economic benefits for both themselves and customers.

Vehicle batteries, if successful, could (ironically enough) eliminate some of the need for bulk storage, and at last energy storage could give significant fuel saving by the use of regenerative braking. It has been estimated that approximately a 10% primary fuel saving could be achieved.

3.5 Dynamic duties of storage

An examination was given above of the possible duties of energy storage in a power system under normal conditions. However, there is also the possibility to use storage methods for operation under transient conditions; so-called dynamic duties.

To maintain sufficient spinning reserve capacity in a purely thermal and nuclear generating system, it is necessary to consume a large amount of fuel and to decrease overall system efficiency. In fact, gas turbines need a time delay of 5–17 min before being synchronised and loaded at their full capacity and therefore can be considered only as standby reserve. Thermal steam generating units and nuclear units are able to increase their output within a few seconds only for 3–5% of their operating load, and consequently only this small fraction can be considered as a spinning reserve.

In order to have the required margin of spinning reserve, it would therefore be necessary to keep a large number of the base-load units in spinning conditions, reducing their output by the same 3–5%, and to cover the demand with generating units having lower efficiency. If the base part of generating structure is completely nuclear, then the particular part where the spinning reserve is allocated has to be replaced for load-covering by conventional thermal generating capacity, with an evident waste of energy and money. The presence of hydro plants can only partly improve the situation; most of the energy storage is capable of being switched rapidly from the charging mode to the discharging mode. This means that any sudden loss of generation in the power system on a sudden increment in load (which was not planned in advance) can be met by energy storage switching provided there is adequate storage rated power and energy capacity.

So if energy storage participates in the supply side of the power system it can perform the spinning reserve duty, avoiding fuel wastage and increasing overall efficiency.

It should not be forgotten that there are other ways that energy storage can perform spinning reserve duty. During a charging period storage can also carry out this duty, since its separation from the system releases an equal quantity of generating power.

Generally speaking, spinning reserve duty is one aspect of the more general frequency control duty. That is the stochastic fluctuations of the difference between generation and load compensating duty. The loss of large generating capacities, to cover the spinning reserve, is the most striking component of such an imbalance, but this appears quite seldom and only in one direction, so that the means allotted to it are also not always able to perform regulation duties for higher frequency fluctuations.

In the case of emergency loss of generation or load, the intervention of the spinning reserve within a few seconds may not be sufficient to maintain so-called dynamic or transient stability so as to avoid, during the first moments of the transition, dangerous frequency drops. In such a case, in order to avoid the collapse of the system, a sudden power input may be necessary, together with a sudden load shedding. This gives a clear advantage to those types of energy storage that are able to make available, instantaneously, their rated power capacity and can therefore be used for improvement of stability, frequency regulation and as a countermeasure against blackout. Incidentally most storage types can perform these duties during the charging regime because it is always possible to drop them off instantaneously.

3.6 Summary of possible applications

The results of the analysis of advantages of energy storage usage generally depend on the characteristics of the supply side of the reference utility, and above all on the form of the load demand curve. As power utilities are evolving, large central generating stations are being built further from load centres. Construction of new generation capacity in urban and suburban areas is increasingly limited by environmental concerns and competition for land use. Any form of modular power plant appears to be ideally suited for urban generation particularly if it can be located at established generating stations as older units are retired or at transmission and distribution substations. In particular, energy storage units, which can be charged at night load times and require no external fuel supply, appear to be well suited for installation in overpopulated urban areas.

It is convenient to group all proposed storage duties according to the required duration of their discharge regimes. One can consider discharge ranges of hours, minutes, seconds and milliseconds.

In the following areas energy storage will work as a buffer compensating any load fluctuations of the range of hours:

- utility load leveling: to improve load factors, reduce pollution in populated urban areas and to make better use of available plants and fuels;

- storage for combined heat and power systems: to improve overall efficiency by offering optimum division between heat and power irrespective of load demands;
- utilisation of renewable energy in its various forms to relieve the burden on finite fossil fuel resources and to improve the environment;
- storage for remote users; and
- storage for EVs: to replace petrol in the long term, reduce urban air pollution and improve utility plant factors.

The minutes range will be covered by storage for industrial mobile power units; to provide better working conditions and as part of uninterruptible power supply systems to improve the reliability of supply, especially in confined areas such as warehouses and mines.

The seconds range will be represented by diesel–wind generators output-smoothing, and storage of necessary energy between pulses of a high-energy particle accelerator.

Millisecond range energy storage units will be used for the improvement of stability, frequency regulation, voltage stabilisation and as a countermeasure against blackout.

There is clearly a limit to the amount of storage plant needed to provide daily smoothing. It is clear that the higher the percentage of the total capacity of the generating park allotted to storage, the smaller the convenience of installing more capacity in storage; both because those means are required to provide longer operating times, with higher costs owing to the larger storage capacities that are needed, and because the advantages to be obtained through the dynamic services become less important, since those services are already provided by the existing means.

The economics of storage plant depends on the mixture of other plant on the system: in particular, whether the proportion of large base-load coal-fired or nuclear plants in the system has grown to more than that is necessary to cover the night-time demand, so that low-cost coal or nuclear-based stored energy becomes available. In this case storage is a complement to a large nuclear programme. Since storage is an essential part of any large-scale renewable programme, it is also considered to be a competitor to a nuclear programme.

The other important aspect is the problem of the choice of the most appropriate characteristics of a storage plant; first, the choice of the rated power and energy storage capacity. The ratio between the storable energy and the unit's rated power must be determined as a compromise between the specific plant cost, which evidently increases with increasing specific storage capacity, and the number and quality of the services it is able to provide; these qualities improve with increasing specific storage capacity.

In general, for the highest peak services it will be convenient to use those methods having specific costs that rise rapidly with storage energy capacity, and others for longer durations.

In order to define the requirements for storage units, it is necessary to carry out some power system analysis on the following topics:

- the different types of energy storage methods in operation at the design stage of the supply side of power utility expansion planning and
- operating experiences and criteria in electricity power systems with storage plants.

Part II

Energy storage techniques

Chapter 4
Thermal energy storage

4.1 General considerations

Direct storage of heat in insulated solids or fluids is possible even at comparatively low temperatures (theoretically from $t > 0\,^{\circ}\mathrm{C}$), but energy can only be recovered effectively as heat. Hot rocks and fireplace bricks have served as primitive heat storage devices from ancient times. This is still the case in industrial furnaces and in the baker's electric oven, where cheap electricity is used to heat the oven during the night.

Thermal energy storage (TES) is efficient due to the high specific melting heat of water. One metric ton of water, just one cubic metre, can store 334 MJ (317 k BTU, 93 kWh or 26.4 ton-hours). No matter what the technology of ice production is (whether ice was produced by modern anhydrous ammonia chillers or hauled in by horse-drawn carts ice was originally transported from mountains to cities for use as a coolant), a sufficiently small storage facility can hold enough ice to cool a large building for a day or a week. Originally the definition of a 'ton' of cooling capacity was the heat to melt 1 ton of ice every 24 h. This definition has since been replaced by less archaic units: 1 ton HVAC capacity equals to 12 000 BTU/h.

TES is ideally suited for applications such as space heating, where low-quality, low-temperature energy is required, but it is also possible to use TES with conventional coal- and nuclear-fired power plants that dominate the installed capacity of electricity utilities and are likely to continue to do so for the near future.

High-temperature thermal storage can be used to utilise heat both in industrial processes and for heat engines. One recent example is the power supply for Stirling engines.

It would be correct to say that the history of steam storage schemes may be traced back as far as 1873. The first displacement accumulator patent was granted in 1893 and the system was developed by Marguerre in 1924 by applying it to a regenerative feedwater heating system using a turbine with overload capacity. It has subsequently been applied to a large power station in the Mannheim municipal works.

Development of a variable-pressure accumulator for power generation began with a German patent given to Dr Ruths in 1913. The first installation was built in Malmo, Sweden, but the largest installation still in operation today was built in 1929 in Germany. This plant is situated at Charlottenburg, Berlin, and operates with 14 bar pressure, 50 MW electric power and 67 MWh storage capacity using a separate peaking turbine.

It is natural to ask why, instead of passing energy through several conversion stages, not just store primary heat from a base-load plant's boiler and recover it when it is most needed?

TES differs from other storage forms for power generation in that energy is extracted in the form of steam between the boiler and the turbo-alternator, as shown in Figure 4.1. Other storage forms are generally charged by extracting the energy as electricity.

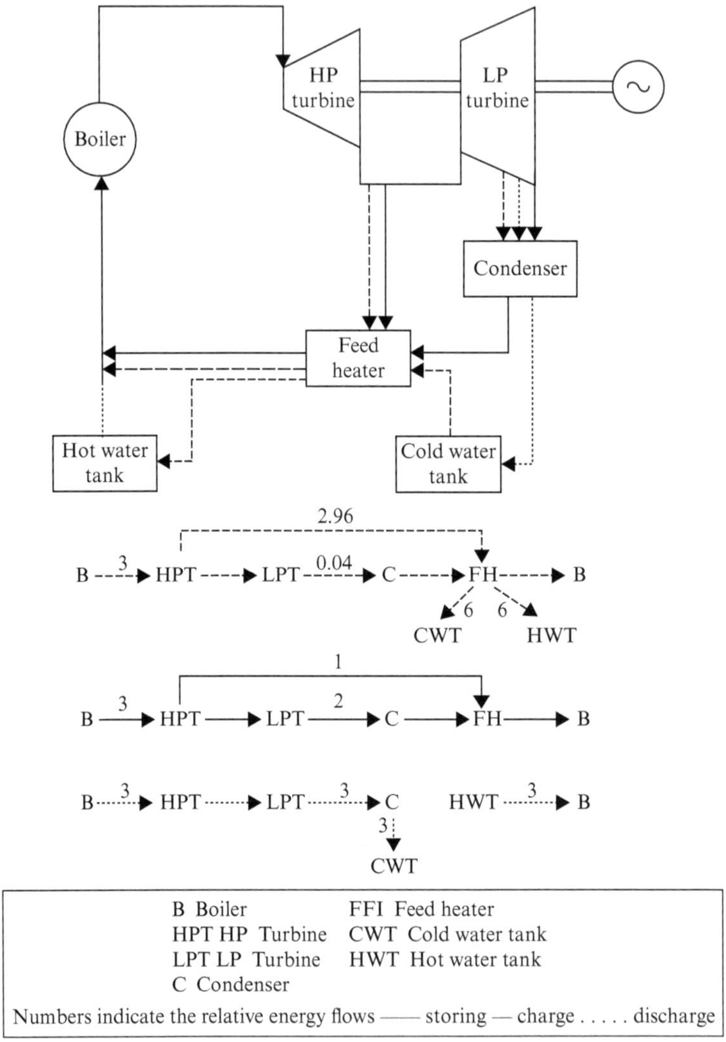

Figure 4.1 One of the number of possible ways in which stored hot water can be used to boost the output of a conventional steam cycle

A power plant used to transfer the heat can be run under constant conditions, independent of electrical demand, since the stored heat can be used to satisfy its fluctuations and the required energy will be readily available to meet load fluctuations swiftly.

Using thermal storage, the boiler can be operated at a constant power level corresponding to the average power output of the base-load plant. The value of storing excess power from a base-load plant during a charging period at night and releasing it during the day has already been shown. Indeed, load cycling of large coal or nuclear plants can be avoided, maximising the return from these expensive plants and providing improved reliability and reduced maintenance.

Thus it is not surprising that this concept was put into practice some 90 years ago at Charlottenburg, where steel vessels ('Ruth accumulators') were used to store a pressurised mixture of power-plant steam and hot water. During peak hours the stored steam was released to drive a turbine-generator set.

In a modern steam cycle, where superheated steam is expanded through a turbo-alternator, approximately 30% is bled off halfway through the turbine and used to preheat water returning to the boiler. The rest continues through the turbine and is condensed in the normal way. So the steam, multiply extracted from the turbines, is used to preheat water returning to the boiler. This is common practice in power plants as it raises the mean temperature of heat reception from the heat source and increases the cycle efficiency.

Heat can be transferred to the store fluid by heat exchangers using steam extracted ahead of and between the turbines, as shown in Figure 4.2, thereby reducing the plant's electrical output. This charging mode is implemented during periods of low electrical demand.

Nuclear- and coal-fired plants are sources of thermal energy. Within them, the thermal energy source can be one of the following (see Figure 4.2):

(i) high-pressure (HP) turbine inlet steam;
(ii) intermediate-pressure (IM) turbine inlet steam;
(iii) low-pressure (LP) turbine inlet steam; and
(iv) intermediate steam-extraction point and feedwater heater (FWH) outputs in the FWH system to raise condensate back to boiler inlet temperature.

The obvious thing to do is to store some energy to increase the flow of steam to the feed heaters at the expense of flow to the condenser, and to use the additional thermal energy of the feed heaters to provide additional hot water, which can itself be stored.

In the TES discharge mode the power output is raised by reducing the extraction of steam for feedheating. The boiler feed temperature is maintained using stored heat, and the steam that is no longer required for feedheating is used to generate additional power either in a separate peaking turbine or in the main turbine, provided there is spare power capacity.

The important point is that the same amount of water will flow through the boiler as before, so the steam flow through the first parts of the turbine remains

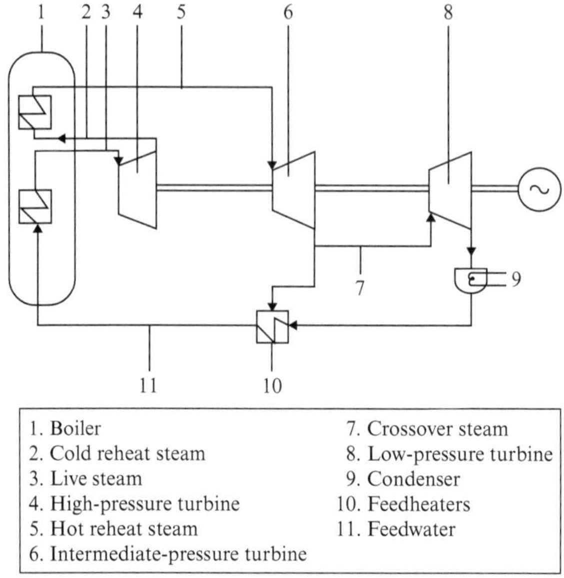

Figure 4.2 Possible TES locations in the thermal part of a conventional power plant

unchanged. The only change as far as the turbine is concerned is the steam flow to the condenser. When it is necessary to produce any additional power, one can use the stored hot water to feed the boiler, cutting off the feed heaters altogether and passing the whole of the steam flow to the condenser. It is in this very simple way that a storage unit can be incorporated into power stations.

The distinctive features of all TES systems are that they have one or more storage media, a form of containment for the storage media, a fluid for heat transfer and heat transport, a source of heat derived from the reference power plant and a means of conversion of stored thermal energy into electricity. There will be periods when the store is neither being charged nor discharged and the plant is then run in a conventional manner.

Air conditioning systems in large building and chilled water systems are the most popular forms of TES application. Air conditioning systems, particularly in commercial buildings, are the main contributors to the peak load demand on hot summer days. In this application a standard chiller may be run at night to produce ice. Water is circulated through the stack of ice during the day to produce chilled water that would normally be the daytime output of the chillers which would be shut off during the peak load hours. Ice storage systems are not too expensive so that storage system may be competitive with conventional air conditioning one.

Large-scale hot water storage can be used for combined heat and power plants and for heating for whole communities. Proposals for the latter have included storage in surface lakes and underground water-holding layers. Surface storage has associated problems such as the interaction with fresh water supply and the risks of chemical and thermal pollution. Hence we will concentrate on underground storage systems, shown in principle in Figure 4.3.

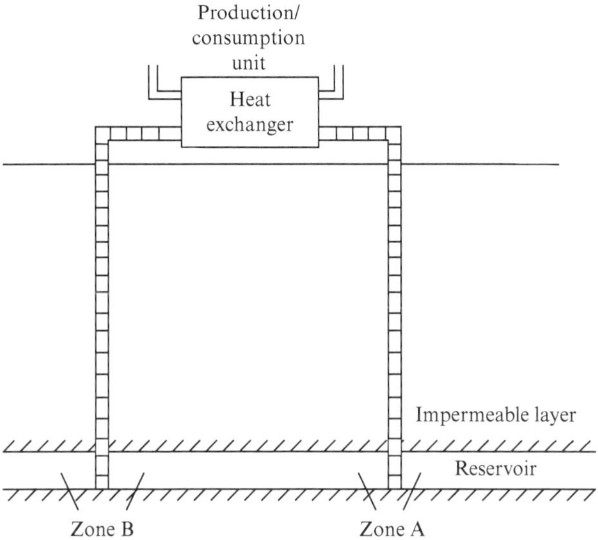

Figure 4.3 Underground layer heat storage

One of the essential parameters for such a storage system is the quantity of water that can be transferred between the underground storage and the surface. The water transfer is determined by the layer thickness $k \times$ the permeability h and the pressure gradient dP. This can be expressed by the flow equation

$$Q - P\frac{kh\,dP}{\mu\ \ln(r_1/r_2)}$$

where Q = amount of water per time unit
P = constant of proportionality
k = permeability
h = layer thickness
dP = pressure difference between outer and inner storage
μ = viscosity of water
r_1, r_2 = radius of the outer and inner storage limits.

If zone B is the storage zone for thermal energy, surplus heat from the production/consumption unit is used in the heat exchanger to heat water from zone A, which is returned heated to the storage stratum in zone B. In periods with heat deficit, the flow is reversed and the store gives off heat.

TES is also used for combustion gas turbine air inlet cooling. Instead of shifting electrical demand to the night, this method shifts generation capacity to the day. The turbine has to be mechanically connected to a large chiller compressor to produce ice at night. At the daytime peak loads, water is circulated between the produced ice and a heat exchanger in front of the turbine air intake, cooling the intake air to near-freezing temperatures. The colder the air, the more the turbine can compress with a given amount of compressor power. Both the turbine efficiency and the electrical generation rise when the inlet cooling system is activated. This method is similar to the compressed air energy storage system that will be considered in Chapter 7.

Small water tanks are widely used for solar heat storage systems, as shown in Figure 4.4. The usual insulation used for individual household heating storage tanks is mineral or glass wool. The insulation has to be sufficient to provide storage times of several months, since load demand and available energy, as mentioned earlier (Figure 2.5), are out of phase. In an optimum design for a single house solar heating system, the size of the storage unit approximately matches the area of the solar panel. It should be mentioned that in temperate climates, the optimum size of the storage unit is larger than in areas where the yearly variations of solar energy are less.

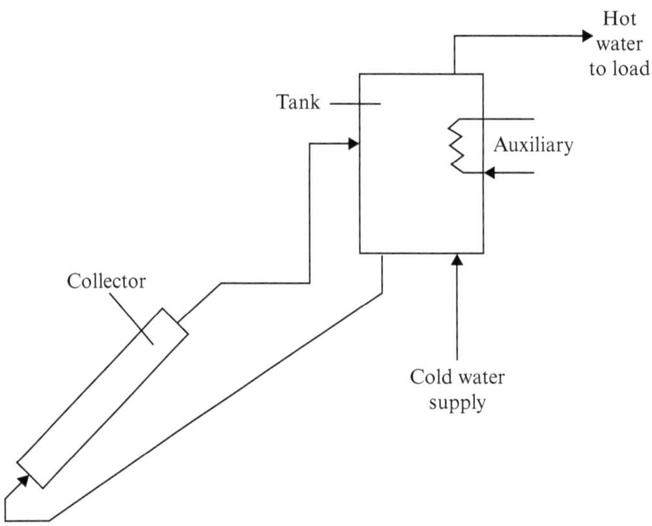

Figure 4.4 Hot water panel system with natural circulation

The barriers to wider adoption of storage heating systems are institutional rather than technical or economic. Neither the required information nor the finance is readily available to encourage investment in such units, and many utilities do not yet offer appropriate rates. For customers who want to buy storage heating systems, however, a variety of electrically heated storage units are commercially available. They include tanks for pressurised hot water, floor-slab heaters and ceramic-brick units for individual room heaters and central building-heating systems. If such systems were to be installed in 10 million homes, the oil and gas saving could be equivalent to 0.5 million barrels of oil per day.

The application of night-time operation of base-load coal and nuclear stations during the summer, when the only storage load is for hot water demand, should perhaps be encouraged as it offers definite advantages in comparison with storage space heating. In the future, water heated with off-peak electricity may also economically replace installations using an oil-fired central heating boiler that cannot be efficient when used at low settings in the summer.

4.2 Storage media

Broadly, there are two TES mechanisms:

(i) sensible heat storage, based on the heat capacity of the storage medium; and
(ii) latent heat storage, based on the energy associated with a change of phase for the storage medium (melting, evaporation or structural change).

Energy can be stored as sensible heat by virtue of a rise in temperature of the storage medium. Water is excellent for this purpose, not only because of its low cost but also because of its high heat capacity (4 180 J/kg/°C). However, with its low melting and boiling points, water is only suitable as a storage medium between 5 °C and 95 °C.

The energy used during a temperature change of, say, 50 °C is of the order of 3.6×10^4 J/kg for rocks, concrete and iron ore. The volume density of the latter is double (2.16×10^5 J/dm^3) that of the former because of the smaller mass density. The values for different working bodies are shown in Table 4.1. Applications for extremely high temperatures (>1 000 °C) have been suggested, but it seems that problems related to materials, such as corrosion, heat shock and other problems associated with heat transfer have been too fundamental to date.

Table 4.1 Comparison of thermal energy storage media

Working body	Type of heat transfer	Energy density		Working temperature (°C)
		10^6 J/kg	10^9 J/m^3	
Water in the steel tank	Heating	0.208	2.08	20–200
Water in the steel tank	Heating	0.54	0.54	350
Water in the steel tank	Heating	2.2	2.2	500
Hot rocks	Heating	0.04	0.1	220–100
Iron	Heating	0.021	0.18	20–350
Ice	Phase-change heat of fusion	0.335	0.33	0
Paraffin	Phase-change heat of fusion	0.17	0.14	55
Salt hydrates	Phase-change heat of fusion	0.2	0.3	30–70
Water	Phase-change evaporation	2.27	2.27	100
Lithium hydride	Phase-change heat of fusion	4.7	4.7	686
Lithium fluoride	Phase change	1.1	2.73	850

High-temperature water (HTW), of adequate quality, has the advantage of being usable directly in the boiler/turbine cycle without interface equipment such as heat exchangers, but requires high-pressure containment for temperatures much above 100 °C and thus is limited in maximum usable temperature unless low-cost pressure containment is available. All the other common storage media considered can be stored at pressures close to atmospheric.

Another large class of storage media is phase-change materials (PCMs). These are materials that melt and freeze at a particular temperature of interest and have a large latent heat of fusion and crystallisation. They have the advantage over sensible heat storage of a higher energy density of storage per degree of temperature change, over the limited temperature range surrounding the fusion point, and can essentially supply heat at constant temperature. When heat is added to, or removed from, these materials, phase change can occur in a variety of ways: melting, evaporation, lattice change or change of crystal-bound water content, where the total energy change is given by the change in enthalpy.

Some inorganic salts, for example, fluorides, have large heat of fusion values but their high melting temperature causes severe corrosion problems. In order to lower the melting temperature, the eutectic mixtures in Table 4.2 have been proposed.

Table 4.2 Melting points for some storage media

Storage medium	Melting point (°C)
Sodium-magnesium fluoride, NaF/MgF_2	832
Lithium-magnesium fluoride, LiF/MgF_2	746
Sodium-calcium-magnesium fluoride, $NaF/CaF_2/MgF_2$	745
Lithium-sodium-magnesium fluoride, $LiF/NaF/MgF_2$	632

The advantages of these fluoride mixtures are that they are chemically stable and can be contained in chromic nickel steel. Their phase-change temperature may be convenient for use as storage for heat engines, but it is much too high for space heating systems.

Salt hydrates have suitable phase-change temperatures for use as storage in heating systems. Their phase transition, however, is often more complex than simple melting, having a solid residue along with a dilute solution.

One of the salt hydrates most often proposed is Glauber salt, $Na_2SO_4.10\,H_2O$. It decomposes at 32 °C to a saturated water solution of NaSO plus an anhydrous residue of Na_2SO_4 while the heat output is approximately 252 kJ/kg. The storage capacity per unit volume over a small temperature range is much greater than for water, suggesting that salt hydrates may be more economical than water storage (since the civil engineering cost is the major expense for a house storage unit).

Metal hydrides have also been suggested for thermal storage. The phase change is the absorption of hydrogen into the lattice of the metal, or metal alloy, during a reaction of the following form:

$$\text{metal} + \text{hydrogen} \longleftrightarrow \text{hydride} + \text{heat}$$

Research, especially in the United States, has proved that a combination of two hydrides in a so-called hybrid storage system can be applied in heating/air conditioning systems. The advantage of this type of heat storage is that there are no losses during storage and the rate of heat re-formation is easy to control. The main objective in hydride research is to find a cheap metal or metal alloy that works at suitable transition temperatures and pressures.

Utilities around the world have recently shown interest in the following high-temperature chemical reaction for TES:

$$CO + 3H_2 \longleftrightarrow CH_4 + H_2O$$

During off-peak hours, heat from one of the primary sources shown in Figure 4.2 is used for heat exchange in the reactor-reformer, where the previously stored methane and water are converted into carbon monoxide and hydrogen and then stored in a separate vessel at ambient temperature conditions. Although the reverse reaction is thermodynamically favoured, it will not occur at these low temperatures and the storage time is in practice infinite. During peak hours the reverse reaction (methanation) is run and the resultant heat is used in the boiler–turbine cycle. Hot air blown through the porous material is usually used as the transfer fluid for high-temperature applications.

The following materials could act as TES central store media using sensible heat:

- high-temperature/low-pressure water ($5 < t < 95\,°C$);
- high-temperature/HP water;
- high-temperature oils;
- molten salts; and
- rocks or minerals requiring oil or molten salt as a heat-transfer medium.

On the other hand, PCMs such as salt eutectics can be used for their latent heat of melting.

4.3 Containment

The main problems of storage design are

- to establish a suitable heat-transfer surface in order to get fast transfer of heat to and from the TES unit; and
- to avoid heat loss to the surroundings so that leakage time is long compared with the required storage time.

The heat loss from a central storage device depends on the surface area of the storage medium container, and the overall storage capacity depends on the volume of the container. The surface area is proportional to the square of the tank's dimensions and, since the volume is proportional to the third power, large TESs need proportionally less insulation than small ones. The stationary temperature distribution $T/T_k = f(x)$, where x is the distance from the centre of a spherical storage volume kept at constant temperature T_k, is shown in Figure 4.5.

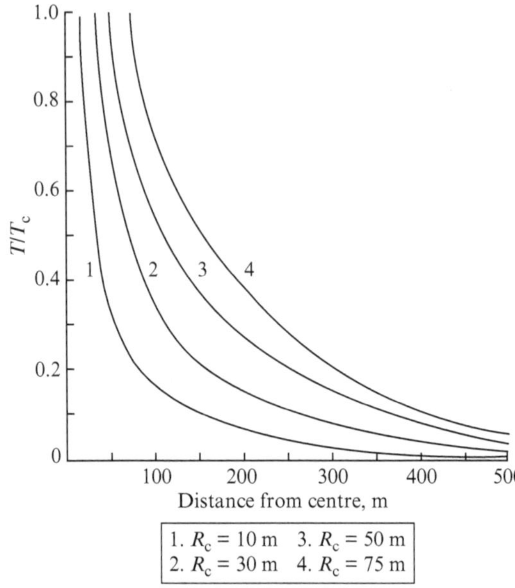

Figure 4.5 Temperature distribution around a special container of different radius R_c with fixed temperature T_c placed in an infinite isotropic medium

The surrounding is considered to be an infinite extent, isotropic medium with a distant temperature of 0 K. T/T_k is given for spheres of radius 10, 30, 50 and 75 m. The curves illustrate the maximum temperature gradients that occur around the container. The small-size, special container clearly results in the largest temperature gradient and therefore relatively in the largest heat leakage to the surroundings.

The overall size of the container is important and during recent years interest has focused upon the possibility of using very large underground reservoirs for long-term storage of heat on a community scale. Such units need relatively little insulation compared with that needed in a small storage unit for a single house.

4.3.1 Steel vessels

Steel vessels are adequate for sensible heat storage in solids and heat-transfer liquids at atmospheric pressure. Storage media can be packed beds of rocks, oils and molten salts. Multiple tanks in modular sizes can be selected for cost and convenience since rather large container volumes are usually required.

For pressure containment typically above 10 bars, steel pressure vessels are readily available, with a background of years of design and operating experience at pressures and/or temperatures above those required for thermal storage. However, storage volumes are far larger than most pressure-vessel applications and cost is the major drawback.

4.3.2 Pre-stressed concrete pressure vessels

Pre-stressed concrete pressure vessels (PCPVs) have been used for primary nuclear reactor containment for over 10 years. However, there has been no specific argument for a TES system using PCPVs as a central store, and none have been built or tested for the pressure and temperature range of interest. PCPVs require cooling to protect the concrete and reinforcing bars from high temperatures; the cooling system is expensive and decreases thermal energy losses. PCPVs, however, are considerably cheaper per cubic metre contained than steel vessels, for comparable duty, and they should be considered for any HTW storage concept requiring pressure containment.

4.3.3 Pre-stressed cast-iron vessels

Pre-stressed cast-iron vessels (PCIVs) were first proposed as a central store for TES in 1974 when a preliminary design was presented by Siempelkamp Giesserei GmbH (Germany). The concept used factory-cast cast-iron arcs, six to a full circle, assembled into multiple cylindrical layers using keyways. External cable wrapping and vertical tendons were used to pre-stress the cast-iron to ensure compression. To contain boiler quality water or HTW, a thin alloy steel liner has to be welded in direct contact with the cast-iron. External thermal insulation is also necessary.

While a small PCIV has been built and conceptual design studies of the application of PCIV to HTW thermal storage have been carried out, no full-scale models for high pressure and temperature have been built. The current concept requires external thermal insulation, part of which must be pressure resistant; also the cast-iron operates hot. Effects of thermal and pressure cycling on the pre-stressing system have not been tested. The advantage of PCIVs is that direct costs per cubic metre of capacity are lower than estimates for PCPV or steel vessels.

4.3.4 Underground cavities

Underground cavities usually take the form of an excavated cavity 30 m or more in diameter in hard rock with a steel liner and concrete stress transfer. A shaft is excavated to such a depth that the overburden sustains the pressure of the storage medium. This technology seems to provide storage at low cost, with additional opportunities from the ability to contain large volumes, and multiple containers sharing a common shaft can provide further cost reduction. Low insulation costs and low 'equilibrium' thermal losses are among the advantages too. The excavation technology for shafts and cavities is well known. A major drawback, however, is that underground cavities in competent rock are limited in number and geographically.

An underground cavity for HTW storage in which the stress in a thin steel liner is minimised by use of compressed air between the liner and the rock, instead of concrete, has been proposed by Margen of Ontario Hydro. The concept has the advantages previously described and in addition the compressed air stress transfer permits external thermal insulation on the tanks. Rock temperatures remain near ambient since the compressed air is cooled. The drawbacks are site selection, which is limited by geology, and leakage of compressed air out, or of ground water into, the cavity may be hard to control by grouting or shot-creting.

4.3.5 Aquifer storage of high-temperature water

Aquifer storage of high-temperature water, that is, porous layers of water-saturated gravel, sand or sandstone confined between impermeable layers, probably has rather low energy-related cost. The doublet well concept permits recycling of hot and cold (or warm) water to and from the same aquifer. The temperature range over which aquifer storage can be effective is still unknown, but a low temperature range of 100–200 °C is definitely usable for feedwater storage. Collins has suggested use at over 300 °C but geochemical effects are uncertain.

An aquifer containment system clearly has a very low energy-related cost for storage per kWh and the capacity for very large amounts of energy to be stored for daily, weekly and even seasonal cycles. However, the power-related cost is significant and includes the cost of drilling and casing the wells, the cost of pumps and pumping energy and the cost of heat exchangers (which are necessary to prevent contamination of the boiler). It is not currently a practical concept, however, in the sense that demonstrations of significant size and useful temperatures have not yet been made.

4.3.6 Summary of containment design

The following options exist for TES containment:

- steel-tank pressure vessels;
- pre-stressed cast-iron vessels (PCIVs);
- pre-stressed concrete pressure vessels (PCPVs);
- underground excavated cavities, steel lined, with high-temperature, high-strength concrete for stress transfer between liner and rock;
- underground excavated cavities with free-standing steel tanks surrounded by compressed air for stress transfer to the rock; and
- underground aquifers of water-saturated sand and gravel confined to impermeable clay layers.

On such a scale it is vital to choose low-cost media for heat storage. Low vapour pressure is obviously essential and so rocks and minerals are usually envisaged. Suitable heat-transfer fluids would then include high-temperature oils for their sensible heat capacity and PCMs, such as molten salts and eutectics, for their latent heat of melting. Storage containment vessels can then be:

- separate hot and cold tanks;
- single tanks in which hot fluid (oil or salt) floats on top of cold fluid, and the boundary between them (the thermocline) moves up and down with the storage discharging and charging cycle; and
- dual-media thermocline tanks in which packed rock beds fill the tank, and oil (or salt) fills the voids and is pumped as a heat-transfer fluid.

4.4 Power extraction

The following means of conversion of stored thermal energy to steam are available:

- flashing high-temperature water to steam and lower-temperature water by throttling the pressure, then passing steam through a peaking or a main turbine;
- using the hot water as boiler inlet feedwater, thus reducing the energy diverted for feedwater heating from the main turbine, increasing its output – a so-called feedwater storage system; and
- using heat exchangers to transfer the energy to cold feedwater, producing either superheated steam or hot feedwater – low vapour-pressure storage media.

There are three main ways of operating the containment vessel or, as it is usually called, accumulator; namely variable pressure, expansion or displacement.

4.4.1 Variable pressure accumulator

The variable pressure or 'Ruths' accumulator mode of operation is shown in Figure 4.6. When fully charged, almost all its volume is filled with saturated hot water, with a small 'cushion' of saturated steam above it. In the discharge mode, steam is drawn off from the top and as the pressure in the steam cushion decreases some of the water in the vessel will flash to steam. All evaporation is internal to the vessel. To charge the accumulator, steam has to be injected and mixed with the water in the vessel.

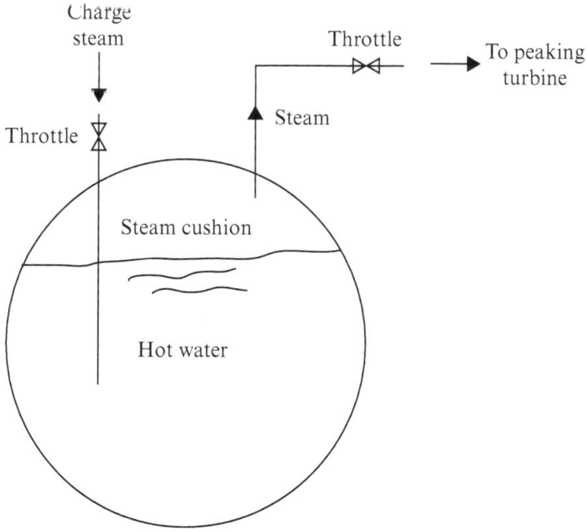

Figure 4.6 Variable-pressure accumulator

4.4.2 Expansion accumulator

The expansion accumulator is shown in Figure 4.7. When fully charged, the accumulator is almost full of hot water with a small steam cushion, as in the variable-pressure mode. As hot water is drawn from the bottom during discharge, enough flashes to steam to fill the tank's volume. This flashing reduces the pressure and temperature of the saturated water and steam slightly compared with the 'Ruths' accumulator. All the water can be removed with a reduction in pressure of approximately 30%. The high-temperature water has to be flashed to steam in evaporators external to the expansion accumulator. During discharge, the water drained from the last flash evaporator has to be collected and stored. The water is, however, at low pressure and temperature, so this cold storage is not expensive. To recharge the expansion accumulator, it is necessary to inject hot water and saturated steam simultaneously.

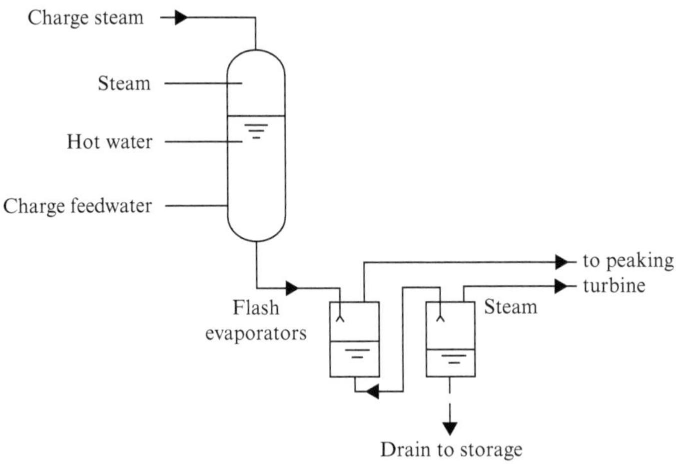

Figure 4.7 Expansion accumulator with flash evaporator

4.4.3 Displacement accumulator

The displacement accumulator is always completely full of water. When fully charged with thermal energy, it contains hot water at the desired temperature; when fully discharged, all the water is cold. As shown in Figure 4.8, hot water is injected at the top during charge and removed from the top during discharge. Cold water leaves and enters at the bottom. Since hot water has lower density than cold, it will float at the top. A sharp temperature gradient or thermocline separates the hot and cold water. Provided mixing currents are avoided it remains stable and sharp and is only limited by the thermal conductivity of water. During discharge, one or more flash evaporators are used to generate steam for peaking turbines. The water drained from the evaporators and the condensate from the turbines are returned to

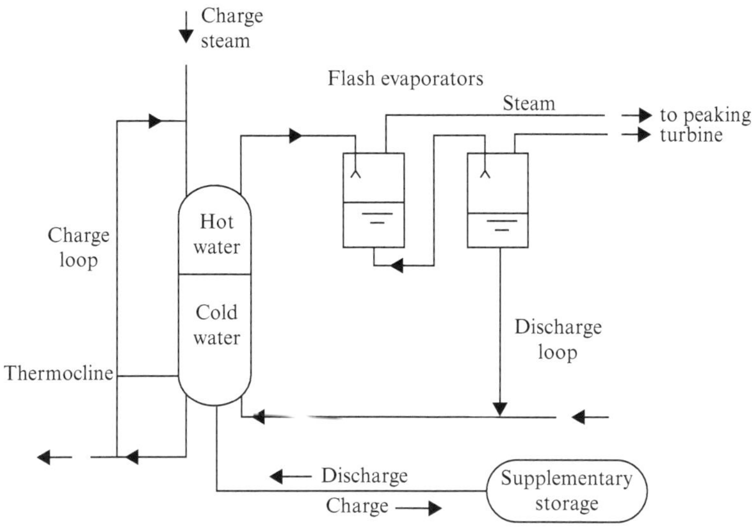

Figure 4.8 Displacement accumulator with flash evaporators

the accumulator as cold water and so no large cold-storage vessel is required. During charge, steam is mixed with cold water taken from the bottom of the tank to raise the temperature to the desired level. Cold water equivalent in mass to the steam is returned to the boiler inlet feedwater to generate more steam. For feed-water storage systems, no steam is needed and the temperature and pressure of the hot water discharged should remain constant unless some steam extraction is used for trimming between storage and the boiler inlet.

The storage medium need not be hot water, but then the stored thermal energy has to be re-transferred to water for steam generation. This requires a heat exchanger. An example of a heat exchanger for atmospheric pressure heat storage is shown in Figure 4.9. This storage unit consists of multiple packed rock beds with hot oil as part of the dual-media system and as the heat-transfer fluid. The containment vessels operate in the displacement mode with a thermocline separating hot and cold oil/rock.

Steam from the chosen heat source can go through three specialised heat exchangers in cascade. The steam entering may be at a temperature considerably higher than the saturation temperature for its pressure, so-called superheated steam. The first heat exchanger is therefore a de-superheater and the condenser then removes the latent heat of vaporisation at constant temperature. The condensate water at saturation temperature may be subcooled in a third heat exchanger to increase the thermal energy stored further and to match the temperature at which the output water is to he charged into the source cycle. On discharge of the storage, condensate from the peaking turbine is heated successively in a preheater, boiler and superheater. Table 4.3 gives a summary of the different concepts of container and steam conversion systems for power plants.

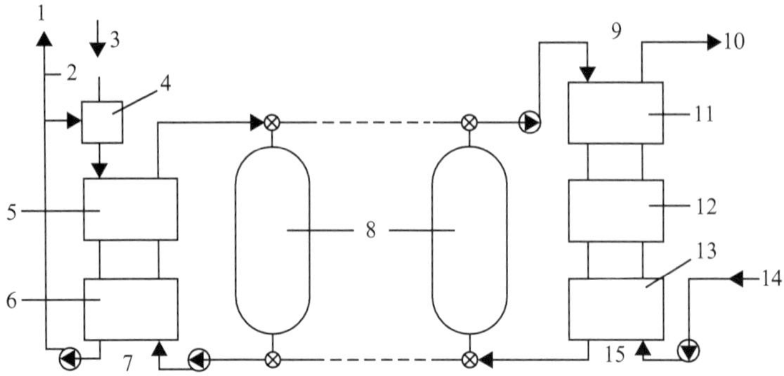

1. Charge	9. Discharge
2. Water to main units	10. Steam to peaking turbine
3. Steam for main units	11. Superheater
4. De-superheater	12. Boiler
5. Condenser	13. Preheater
6. Subcooler	14. Water from peaking turbine
7. Storage heater	15. Steam generator
8. Oil/rock thermal energy storage tanks	

Figure 4.9 Heat exchangers for sensible-heat-storage scheme

Table 4.3 Preferred selection of containers and steam conversion systems for power generation

Containment	Steam conversion system
Steel vessel	Displacement accumulator
PCIV	Expansion accumulator with one evaporator
PCPV	Variable-pressure accumulator
Underground cavern using concrete stress transfer	Variable-pressure accumulator
Underground cavern using air stress transfer	Displacement accumulator with either feedwater storage or three evaporators
Aquifer	Feedwater storage
Oil/feedwater tank	Hot-and-cold tanks
Oil/packed rock bed/thermocline	Steam generator for peaking turbine
Molten salt	Steam generator for peaking turbine

4.5 Thermal energy storage in a power plant

There are two major variants of steam conversion to electrical energy. The first is the use of an oversized turbogenerator designed for base-load and peaking flow rates. The second is the use of a separate peaking turbine for increased thermal capability. The latter option is technically preferable as it offers greater operational flexibility, giving a wide range of additional capacity, and it offers improved availability provided the peaking turbine can be powered directly from the boiler as well as from the TES unit.

Figure 4.2 shows the possibilities for diverting steam for storage ultimately to be used in a peaking turbine. It should be mentioned that a TES unit using hot water as a storage medium stores the water at saturation pressure. Consequently, significantly superheated charging steam, as is available in a conventional coal-fired steam cycle, has to be de-superheated and condensed before storage. This leads to a significant loss of available energy. For this reason, hot reheat steam from a coal-fired boiler appears to be a less attractive option than cold reheat steam for TES charging. The cold reheat steam (or live steam) application, however, implies variable steam-flow ratios between superheater and reheater, and leads to excess reheater tube temperature and increased problems of forced outage. To overcome this difficulty, new boiler designs and elimination of the reheater from the thermal cycle will be needed. Container type, choice of medium and flow arrangement have a significant bearing on the losses of energy through the TES unit. Let us choose the turnaround efficiency as a primary measure of performance for a thermal plant with a TES unit. We define this efficiency as a ratio of the electrical energy generated by the thermal power plant during the TES discharge regime in peak hours to the reduction of electrical energy generation during the charge regime in a trough. The efficiency is given as follows:

$$\xi_s = \frac{N_p/P - 1}{1 - N_h/Pt_c} \frac{t_d}{t_c}$$

where N_p, N_h, P = power generation during discharge, charge and normal operation, respectively

$$\frac{t_d}{t_c} = \text{discharge to charge time ratio}$$

The net overall efficiency (ξ_{net}) evaluated over a 24-h period, defined as the ratio of the electrical energy generated to the thermal energy output from the boiler, will be given as follows:

$$\xi_{net} = \xi_{base} - \left(\frac{t_c}{24}\right)\left(1 - \frac{Nh}{P}\right)\xi_{base}(1 - \xi_s)$$

where ξ_{base} is the efficiency of the thermal power plant in normal operation.

It is obvious that during the TES charge regime the main turbine operates at a reduced load N_h, which is constrained by the minimum tolerable steam-flow rate, which ensures that the main turbine heat rate is not significantly worsened. In this respect, the ratio N_h/P is nominally less than 0.7; it typically carries a significant efficiency penalty and should be avoided.

It could be concluded from the net overall efficiency equation that short charging times are desirable for a high overall ξ_{net}. However, long charge times t_c are needed to ensure that the turbine operates as close as possible to its nominal output during the TES charge regime. Therefore, the main task is to minimise the term $(1 - N_h/P)(1 - \xi_s)$, which dominates. This trend is particularly important for large peaking-power swings.

As an example of the relatively small impact that TES for load levelling might have on net efficiency, let us consider a load curve with a power swing equal to 0.3 and a charge–discharge ratio $t_c/t_d = 1$. This would have a net efficiency of 0·98 relative to full-load base efficiency, which corresponds to a TES turnaround efficiency of approximately 0.8.

The discharge time t_d is described as 6 h, which corresponds to daily regulation. Operation at progressively higher discharge times, which means weekly or even seasonal regulation, increases the containment volume needed and hence the central store cost, and makes it impossible to charge the TES for long periods in large power-swing situations, thus reducing the maximum attainable turnaround efficiency.

Other parameters with an influence on the variation of TES turnaround efficiency are the storage and throttle pressures, as shown in Table 4.4.

The cold reheat steam pressure in a coal-fired thermal plant is typically 43 bar, whilst in a PWR plant the highest steam pressure available is nominally 67 bar; both are suitable for a TES unit. Reducing the storage pressure reduces the cost of central storage but increases throttling losses if the source of charge steam remains the same. Low storage pressures are therefore not an attractive option unless charged with crossover steam at typically 10 bar, as shown in Figure 4.2. Low-pressure storage, however, dictates a very large peaking turbine and condenser and, furthermore, for large power swings the turnaround efficiency is penalised because the swing must be accomplished in the low-pressure turbine alone, causing it to be operated far from the optimal regime.

The last source for TES charge shown in Figure 4.2 is the feedwater train. A displacement accumulator or a hot-and-cold two-tank system is suitable for feedwater storage. The 'cold' tank should ideally be near to 100 °C with a steam cushion. No steam turbine is currently capable of operating with all the external steam cut-off, which would give the maximum power swing. Therefore, in order to handle the extra steam flow, either a peaking turbine or the exhaust area of the main turbine in both coal and PWR plant must be increased by 25%. The maximum peaking swing for both plants even then is limited to approximately 17%, but both plants achieve high specific output and high turnaround efficiency – approaching 0.9 for an 8-h charge, 6-h discharge (daily regulation).

Atmospheric pressure TES units usually exploit low vapour-pressure fluids and heat exchangers, as shown in Figure 4.9. The values of the temperature at all heat exchanger pinch points, and the ratio of the quantities of heat storage fluid and charge steam involved, are the key parameters that define thermodynamic performance. Properties and flow rates of generated steam can then be calculated.

Stored energy can generally be used in the same ways as for hot water systems. The feedwater heating system (where power swings are limited to approximately 0.17)

Table 4.4 Sources of loss and efficiencies in TES systems

TES site in the thermal cycle	Power transformation system	Reasons for stored losses of energy		Turnaround efficiency
		Charge regime	Discharge regime	
Internal steam generation	Ruths accumulator	Pipe friction Throttling	Throttling peak turbine efficiency difference compared with base turbine	0.7–0.8
External steam generation	Displacement accumulator	Heat transfer and mixing	Heat transfer	0.6–0.7
		Energy for recirculation	Throttling energy for recirculation Peak turbine efficiency difference	
Expansion accumulator	Heat transfer	Heat transfer		0.6–0.7
	Energy for Pumping	Throttling Peak turbine efficiency difference		
Feedwater storage	Displacement accumulator used directly	Energy for recirculation	Energy for recirculation Peak turbine efficiency difference	0.8–0.9
Indirect use with oil or rocks	Heat transfer	Heat transfer		0.5–0.6
	Energy for Pumping	Energy for pumping Peak turbine efficiency difference		

is the simplest configuration involving the smallest number of heat exchangers. For larger power swings, up to 0.5, a packed-bed thermocline storage system can be used, acting as a steam generator (see Table 4.3). Assuming a heat exchanger efficiency of 0.93, a temperature approaching 11 °C and with, for example, a 25% void in a gravel bed filled with oil, this storage device when coupled to a 1140 MW base-load plant would typically require the characteristics given in Table 4.5. The oil/feedwater heat exchanger is larger and therefore more expensive than the charge unit, since it is a sensible heat rather than latent-heat exchanger and can accommodate a higher flow rate assuming an 8-h charge, 6-h discharge cycle. The cost of the central store components (oil, gravel and container) can be modest in comparison with heat exchangers – cost ratios of 1:4 have been reported.

Table 4.5 Typical values for parameters of oil/packed-bed storage systems

Discharge FWH	1.65×10^{5}
Charge extraction heaters (m^3)	3.0×10^{4}
Containment volume (m^3)	6.35×10^{4}
Oil quantity (gal)	3.8×10^{6}
Gravel quantity (tons)	1.36×10^{5}

4.6 Economic evaluation

The energy-related costs of different storage media and different forms of containment are influential in the overall costs of TES. Usually, low-cost storage media require expensive forms of containment, and vice versa. The only exception to this rule is aquifer containment. Table 4.3 introduced TES systems with different mechanisms of containing high-temperature water under pressure. The cost of containment is a function of both design pressure and volume. The cost against volume relationship is not usually linear for a single pressure containment, but for volumes many times larger than the largest practical unit size a linear relationship is a reasonable approximation. From studies in the United States, underground cavity costs are considerably lower than for other containment forms, especially if the shaft for access to the cavities is regarded as part of the power transformation system.

Power-related costs for systems are usually dominated by the peak turbine. In addition, for low vapour-pressure systems, both the power cost and energy-related cost are sensitive to the heat exchanger pinch point and the ratio of the flow of heat-transfer fluid to charge steam. For example, a decrease in the value of the pinch point will increase the (energy-related) cost of the heat exchanger but improve the turnaround efficiency, raising the pressure and temperature of the discharge steam, and thereby reducing the power-related cost. Counteracting cost trends such as these creates a competitive effect and leads to a classical optimisation problem for minimising the cost of TES.

Table 4.6 gives relative cost estimates for various TES systems, as well as presenting the energy- and power-related costs of economically attractive concepts linked with a coal cycle. As previously discussed, steam reheat is omitted for 'steam generator' storage-linked coal cycles. Alterations are then needed in the main turbine to cope with the higher moisture levels in the working steam. TES could be used in a PWR cycle; the steam–steam reheater could be retained as it is separate from the steam generator.

Implicit in the feedwater storage cycle power-related cost in Table 4.6 is the special design of the main turbine to cope with variable steam flows during TES charge and discharge regimes. Reduction in this cost may be achieved by introducing a peak turbine for the excess discharge steam flow, as shown in Figure 4.9, and choosing a cycle arrangement that minimises non-nominal running of the main turbine during the TES charge regime.

Non-economic factors may determine whether TES or other storage systems are preferable for load-following. For example, TES differs from other storage methods as it is not a separate storage plant elsewhere in the power system but is a component of the thermal power plant itself. This can mean a loss of availability of storage and greatly affects plant siting as well as operating flexibility, reliability, safety and environmental acceptability. For example, two of the storage selections in Table 4.6 are only available in suitable geological areas.

Table 4.6 Relative cost estimates for selected TES concepts

TES concept	Relative estimate cost			Coal plant with TES turnaround efficiency
	C_e	C_p	C_t	
Under concrete stress transfer variable accumulator	0.19	0.50	0.69	0.80
Oil-packed rock bed steam generator	0.19	0.56	0.75	0.66
Underground compressed air stress transfer evaporation	0.11	0.67	0.78	0.88
Salt-rock aquifer			0.90	
Feedwater storage based on:				
PCPV			1.0	
Oil-rock			1.0	
PCIV			1.3	
Steel			1.4	
Phase-change materials			1.5	

Operating flexibility can be less attractive for small peaking increments such as 5–20% of the base-load capacity compared with increments of 30–50%. A conventional feedwater storage unit with its limited available power swing may therefore be less attractive than other systems. Operating flexibility is also concerned with the duration of discharge at full capacity that is available. The energy-related cost is proportional to the time of discharge, whereas the power-related cost is not.

PCIV storage is clearly a high-cost kWh system, whereas TES with aquifer containment, although not an attractive proposition for short discharge design, is possibly the most suitable TES system for weekly or seasonal storage.

Safety aspects and operating hazards need to be thoroughly investigated when ensuring that TES does not decrease the reliability of the components of a power plant. For example, a clean TES medium is essential to avoid boiler feedwater contamination since storage media that are not specially prepared passing through the turbine could cause blade erosion at least. Protection against the possibility of incursion of oil increases heat exchanger costs and significantly reduces the turnaround efficiency of such plant.

If the storage options in Table 4.6 were arranged in order of ready availability, the most expensive steel feedwater TES would be most applicable as it has least technical difficulties.

The decision whether to use TES units for future power generation is strongly influenced by practical considerations; these will also be considered in the discussion of other forms of storage in the following chapters.

Chapter 5
Flywheel storage

5.1 General considerations

Storing energy in the form of mechanical kinetic energy (for comparatively short periods of time) in flywheels has been known for centuries, and is now being considered again for a much wider field of utilisation, competing with electrochemical batteries.

In inertial energy storage systems, energy is stored in the rotating mass of a flywheel. In ancient potteries, a kick at the lower wheel of the rotating table was the energy input to maintain rotation. The rotating mass stored the short energy input so that rotation could be maintained at a fairly constant rate. Flywheels have been applied in steam and combustion engines for the same purpose since the time of their invention. The application of flywheels for longer storage times is much more recent and has been made possible by developments in materials science and bearing technology.

The energy capacity of flywheels, with respect to their weight and cost, has to date been very low, and their utilisation was mainly linked to the unique possibility of being able to deliver very high power for very short periods (mainly for special machine tools).

The energy content of a rotating mechanical system is

$$W = 0.5\, I\omega^2$$

where I = moment of inertia
ω = angular velocity

The moment of inertia is determined by the mass and the shape of the flywheel, and is defined as

$$I = \int x^2 dm_x$$

where x is the distance from the axis of rotation to the differential mass dm_x. Let us consider a flywheel of radius r in which the mass is concentrated in the rim. The solution of the integral will then be simple since $x = r =$ constant:

$$I = x^2 \int dm_x = mr^2$$

and

$$W = 0.5\, r^2 m\omega^2$$

The last equation shows that the energy stored depends on the total mass of the flywheel to the first power and the angular velocity (the number of revolutions per time unit) to the second power. This means that, in order to obtain high stored energy, high angular velocity is much more important than the total mass of the rotating flywheel storage.

The energy density W_m (the amount of energy per kg) can be derived directly by dividing the energy content of the flywheel by its mass:

$$W_m = 0.5\, r^2 \omega^2$$

The volume energy density W_{vol} is derived by substituting m in the equation with m expressed as the mass density ρ multiplied by the volume:

$$W_{vol} = 0.5\, \rho r^2 \omega^2$$

The tensile strength of the material dictates the upper limit of angular velocity. In our example, the tensile stress σ in the rim is given by

$$\sigma = \rho \omega^2 r^2$$

so that the maximum kinetic energy per unit volume will be

$$W_{vol\max} = 0.5\, \sigma_{\max}$$

Thus, if the dimensions of the flywheels are fixed, the main requirement is high tensile strength. High values of W_{vol} are important in certain cases, but if transport applications are involved it would clearly be better to use the maximum W_{m} as a criterion. By combining mass and volume density equations as a function of ρ and σ, it is possible to define maximum mass density as follows:

$$W_{m\max} = \frac{0.5\, \sigma_{\max}}{\rho}$$

An elementary theoretical analysis of the flywheel therefore shows that the energy storable, per unit mass, is proportional to the allowable tensile stress divided by the material density. Contrary to most people's intuition, the maximum energy storage capacity of a flywheel, for a given load on its bearings, is achieved with a flywheel not made of a heavy metal, but of a material that combines low density with high tensile strength.

5.2 The flywheel as a central store

The factor 0.5 in the energy density expression relates to a simple rim flywheel. The general expression for any flywheel, made from material of uniform mass density, may be given as follows:

$$W_{m\max} = \frac{K\sigma_{\max}}{\rho}$$

The value of K depends on the geometry of the flywheel, and is usually called the flywheel shape factor or form factor. Essentially, the value of K comes from the expression for the moment of inertia I. Values of K for a number of flywheel shapes are given in Table 5.1.

Table 5.1 Flywheel shape factors

Flywheel shape	***K***
Constant stress disc	0.931
Flat unpierced disc	0.606
Thin rim	0.500
Rod or circular brush	0.333
Flat pierced disc (od/id = 1.1)	0.305

Object	***K* (varies with shape)**	**Mass kg**	**Diameter**	**Angular velocity rpm**	**Energy stored**	**Energy stored**
Bicycle wheel at 20 km/h	1	1	700 mm	150	15 J	4×10^{-3} Wh
Bicycle wheel, double speed (40 km/h)	1	1	700 mm	300	60 J	16×10^{-3} Wh
Bicycle wheel, double mass (20 km/h)	1	2	700 mm	150	30 J	8×10^{-3} Wh
Concrete car wheel (19 km/h)	1/2	245	500 mm	200	1.68 kJ	0.47 Wh
Train wheel at 60 km/h	1/2	942	1 m	318	65 kJ	18 Wh
Giant dump truck wheel at 30 km/h (18 mph)	1/2	1 000	2 m	79	17 kJ	4.8 Wh
Small flywheel battery	1/2	100	600 mm	20 000	9.8 MJ	2.7 kWh
Regenerative braking flywheel for trains	1/2	3 000	500 mm	8 000	33 MJ	9.1 kWh
Electrical power backup flywheel	1/2	600	500 mm	30 000	92 MJ	26 kWh

The shape factor K is a measure of the efficiency with which the flywheel's geometry uses the material's strength. The ideal shape would be a constant stress disc where all the material is uniformly stressed biaxially so that the tangential and radial stress components remain at equal levels to an outer radius of infinity and therefore $K=1$. It should be mentioned that this is not a particular shape for a solid flywheel.

In order to obtain maximum energy storage density, a special design has been proposed, where maximum stress is obtained throughout the flywheel. Such flywheels are thickest near the axis and thinnest near the rim. The shape factor of these truncated conical discs is about 0.8. This modern design is in contrast to the classical steam engine design.

Other designs have to be applied to flywheels made of fibre materials, where the tensile strength is high in only one direction. An efficient solid disc construction is not possible with composite materials, due to the poor transverse strength of these materials. The radial component of the biaxial stresses developed when the disc is spun up tends to cause early failure. Only at the expense of reduced tangential strength is it possible to build up strength with fibres in the radial direction.

An alternative flywheel design for composite material is a radially thin hoop-wound rim. An ideally thin rim has all the material in hoop stress, thereby making full use of the tangential stress capability. In this case the shape factor will be $K=0.5$. Rims of finite thickness are used in practice and the material is not stressed uniformly so the shape factor is typically $K=0.4$. Rods of aligned fibres mounted in a 'sweep's brush' configuration represent another construction, in which case the stress varies along the rods and the shape factor becomes $K=0.3$.

For anisotropic materials with a given ratio of hoop to radial modulus, it may be shown that the inner/outer radius ratio may be set such that radial delamination occurs just prior to the ultimate tangential stress being achieved. The need to build a massive chamber to contain the heavy splinters that would be produced if a solid metal flywheel fractured is avoided since the delamination failure takes the form of a pulverising process. Therefore, the orthotropic nature of composite materials should allow a safer flywheel to be developed. Flywheel systems rotating in opposite directions have been proposed in order to avoid the gyro effect, which is proportional to the first power of the angular velocity.

Construction of neither the rim nor the rod is volume-efficient since the available energy is stored in a small fraction of the swept volume. It has been proposed to have a number of different sized rims fitted closely inside one another within the rim construction. With this design it is necessary to change the mass, speed or modulus of the inner rings to avoid large gaps being opened up between the rings as the flywheel is spun up. However, this form of construction creates serious problems for a power transformation system.

The amount of energy that can be stored by a flywheel is determined by the material design stress, material density and total mass, as well as flywheel shape factor K. It is not directly dependent on size or angular speed since one of these can be chosen independently to achieve the required design stress. Material properties also govern flywheel design and therefore allowable K values. In order to take maximum advantage of the best properties of highly anisotropic materials, the flywheel shape is such that lower *K* values have to be accepted compared with those normally associated with flywheels made from isotropic material.

Details of flywheel energy storage and relative costs are given in Table 5.2 for a number of both anisotropic and isotropic materials (mild steel, maraging steel, titanium alloy). The design stresses for the anisotropic materials are based on the tension/tension fatigue properties at about 10^5 cycles where the load is applied along the fibre direction; a hoop-wound rim construction is assumed and the thickness of the rim is such that no radial delamination occurs below the design hoop stress. The design stress for the isotropic materials is also based on the tension/tension fatigue properties at 10^5 cycles. A solid disc construction is assumed for these materials; however, consideration is also given to the marked biaxial loading that arises and the resulting stress distribution throughout the disc.

Table 5.2 Comparison of flywheel energy storage

Material	Design stress 10^6 N/m	Density 10^3 kg/m^3	Useful energy 10^3 J/kg	Mass of the flywheel 10^3 kg	Relative cost for material and fabrication pu/J
Wood birch	30	0.55	21.0	1 720	1.0
Mild steel	300	7.80	29.5	1 220	1.11
E-glass 60% fibre/epoxy	250	1.90	50.4	713	0.523
S-glass 60% fibre/epoxy	350	1.90	70.5	509	0.492
Maraging steel	900	8.00	86.4	417	2.18
Titanium alloy	650	4.50	110.8	325	6.98
Carbon 60% fibre/epoxy	750	1.55	185.4	194	0.34
Kevlar 60% fibre/epoxy	1 000	1.40	274.3	131	0.26

Material costs given for anisotropic materials are manufacturers' projected prices for large quantities and the fabrication costs are for large volume production. The material and fabrication costs for isotropic materials are those currently quoted by manufacturers for larger forgings or fabrications. It is clear from Table 5.2 that flywheels made of anisotropic fibre composite material are considerably cheaper to construct (material and fabrication cost) than wood or metallic alloy flywheels. Recently developed fibre-reinforced plastics are suitable materials. It should be mentioned that only since their introduction has it become possible to propose the use of flywheels capable of storing energies as great as 3.6×10^{10} J for recovery over a period of a few hours.

Glass fibres offer better possibilities than carbon. Glass fibres have practically the same intrinsic strength as carbon fibres, but they have at least two drawbacks. First, they tend to be more elastic than the normal bonding materials so that the bonding material cracks under high strain. Second, the glass suffers from stress corrosion on exposure to moisture (which may gain access to the surface as the bonding cracks) and this greatly reduces the ultimate tensile strength of the fibres.

Professor Richard Post of the University of California has offered solutions to these difficulties. He suggested that the composite should be made in an integrated factory so that the glass can be prepared, bonded and sealed under dry conditions. To reduce stress corrosion, which proceeds at a very temperature-dependent rate, he suggested operating the flywheel at deep-freeze temperature, and proposed an effort to develop bonding materials of much greater elasticity.

Developments in glass–fibre/epoxy design stress are very promising but as far as carbon–fibre/epoxy and Kevlar–fibre/epoxy are concerned it does not seem likely that their design stress will be uprated. However, it has been shown that it is possible to apply radial compression to composite material flywheels, so that the centrifugal loading must first neutralise this compressive stress before going tensile to a fatigue-safe tensile limit. The fatigue stressing is now compression/tension and significant increases in energy storage arise. The transverse strength of fibre composite material flywheels may be increased by copolymerising the epoxy matrix with elastomer. Hybridisation with cheap glass fibres could cut the cost of flywheels substantially. This is a rapidly developing field and further cost reductions are foreseen.

Wood does not seem attractive as a flywheel material. Table 5.2 shows that the mass would be enormous, and because of the low density its required volume would be a crucial limitation factor. Design stress could probably be uprated, but even then there will still be the difficulty of obtaining large pieces of wood having identical properties.

Maraging steel and titanium are not economically attractive as flywheel materials and there seems no chance of changing this position in the foreseeable future. However, for mild steel (where the fabrication cost is almost ten times greater than the material cost), some reduction in total cost may occur on a large volume production basis. Moreover, a storage capacity of 7.2×10^4 J/kg has been quoted for a pre-stressed laminated steel flywheel, more than double the energy storage given in Table 5.2.

5.3 The energy discharge problem

Not all the stored energy can be used during discharging. The useful energy per mass unit can be given as follows:

$$\frac{E}{m} = \frac{(1 - s^2)K\sigma}{\rho}$$

where s is the ratio of minimum to maximum operating speed, usually taken to be 0.2.

5.4 Applications of flywheel energy storage

Some of the problems of charging and discharging energy from flywheels have been familiar for many years. Flywheels have been used for short-term energy storage applications so as to maintain steady rotational speeds in machines. Millisecond duration pulses of high power for the magnets of nuclear particle accelerators, for example, may be discharged from a flywheel energy storage system in which the energy has been charged over long periods (hours duration). Navy and spacecraft control devices have also used flywheels.

Most present-day development work deals with the manufacture of small-scale flywheel energy storage systems for use in transportation applications as 'on the road' energy sources, storing energy over short time spans and as static power sources. Only very recently has flywheel energy storage been considered for use in electricity power systems. It seems reasonable, as experience is gained in the design, manufacture and modes of operation of small-scale flywheels, that it should be possible to consider medium- and large-scale versions able to discharge energy for several hours safely and reliably.

Electricity distribution systems require flywheel energy storage systems with an energy capacity of 3.6×10^9 J. The relevant mass required to provide this capacity is shown for a number of promising materials in Table 5.2. The high specific design stresses of Kevlar–fibre/epoxy and carbon–fibre/epoxy allow storage of a large amount of energy in a relatively light flywheel. The minimum energy capacity of 3.6×10^9 J required by a power system could be achieved by multiring flywheel energy storage made of hoop-wound Kevlar–fibre/epoxy material. The angular speed of this flywheel would be about 3 000 rev/min. The diameter of the largest ring would be about 5 m, the length about 5 m and the total mass of the flywheel energy storage would be 130×10^3 kg.

The turnaround efficiency of a flywheel energy storage system during the charge–store–discharge period will depend on the duration of the keeping store regime. There are two main sources of losses in the flywheel: windage and bearing. Windage losses can be reduced to a low level by running the flywheel in a vacuum chamber. Bearing losses for a typical 200 t rotor have been estimated at 2×10^5 J/s. Based on this figure, the turnaround efficiency ξ_s should be 85% for discharge immediately after charge, but will fall to 78% after 5 h and 45% after 24 h of the keeping regime.

An example of a large conservative design flywheel is the steel flywheel energy storage supplying short pulses of energy to the Stetallarator at Garching. The required energy of 5.33×10^{12} J can be supplied by the flywheel with a power transformation system rated at 150 MW in a 10 second discharge regime. The flywheel weight is 223×10^3 kg and it has an energy density about 1.58×10^4 J/kg.

The bearing design will constrain the weight of a practical rotor to 200 t maximum and hence a single composite material flywheel energy storage unit would almost certainly be limited to 3.6×10^9 J energy capacity.

Maximum storage capacity will be achieved at the point of mechanical failure of the flywheel. It is clear that, for safety reasons, the stored energy must be kept well below this limit since a fault could cause a sudden release of stored energy with an explosive effect.

Finally, it should be noted that realistic cost comparisons of different energy storage schemes should be made on a total-life basis for the unit. The time factor may affect considerably the conclusions reached from simple capital cost comparisons and, in this respect, flywheels can be competitive against other energy storage techniques such as batteries.

Chapter 6
Pumped hydro storage

6.1 General considerations

Hydro power became cheaper than gas for public lighting as early as 1881 when the world's first public electricity supply system was built for Godalming in Surrey, UK. As soon as hydroelectric power systems became widespread associated developments for energy storage, using pumping water, followed in Italy and Switzerland in the 1890s. We have more than a century of experience in the storage of natural inflows as an essential feature of the exploitation of potential hydraulic energy. Reservoirs can be used to store artificial inflows obtained by utilising the energy available in power systems during demand troughs from generating capacity that is not fully loaded.

At the start of the last century, all hydroelectric plants with reservoirs were equipped with certain pumping mechanisms in order to supplement the natural inflow to the upper reservoir; the main idea was to create seasonal storage in a hydroelectric power system.

Only in the second phase of evolution, and predominantly in thermoelectric systems, were special hydroelectric plants built with inflows from pumping alone. These so-called pure pumped-storage plants were designed mainly for daily or weekly storage.

Pumped hydro storage is the only large energy storage technique widely used in power systems. For decades, utilities have used pumped hydro storage as an economical way to utilise off-peak energy, by pumping water to a reservoir at a higher level. During peak load periods the stored water is discharged through the reversible pump-turbines to generate electricity to meet the peak demand. Thus, the main idea is conceptually simple. Energy is stored as hydraulic potential energy by pumping water from a lower level to a higher level reservoir. When discharge of the energy is required, the water is returned to the lower reservoir through turbines that drive electricity generators.

Pumped hydro storage usually comprises the following parts: an upper reservoir, waterways, a pump, a turbine, a motor, a generator and a lower reservoir, shown schematically in Figure 6.1.

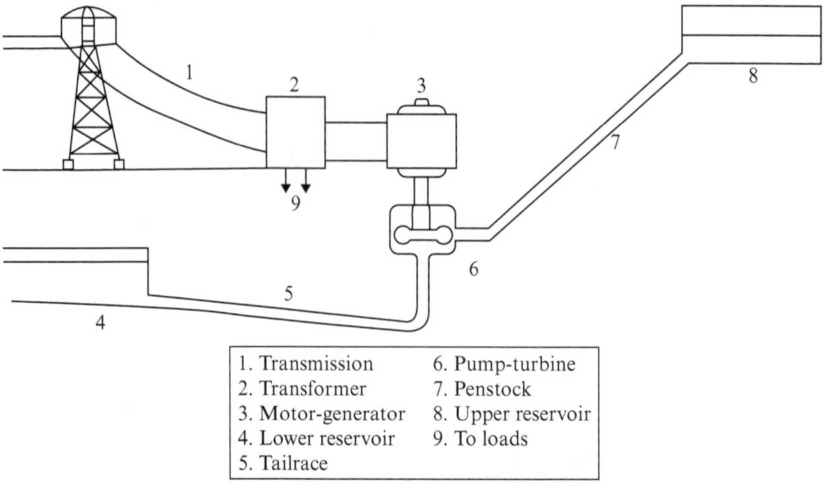

Figure 6.1 Pumped hydroelectric energy storage

As in any hydraulic system, in hydro pumped storage there are losses during operation, such as frictional losses, turbulence and viscous drag, and the turbine itself is not 100% efficient. The water retains some kinetic energy even when it enters the tailrace. For the final conversion of hydro power to electricity, we also have to account for losses in the generator. Therefore, we will define the overall efficiency of hydro pumped storage ξ_s as the ratio of the energy supplied to the consumer while generating, E_g, and the energy consumed while pumping, ξ_p. It is clear that this efficiency depends on pumping, ξ_p, and regeneration, so that

$$\frac{\xi h = E_g}{E_p = \xi_p \xi_g}$$

The energy used for pumping a volume V of water up to height h with a pumping efficiency ξ_p is

$$E_p = \frac{\rho g h V}{\xi_p}$$

where ρ = mass density.

The energy supplied to the grid while generating with generating efficiency ξ_g is given by

$$E_g = \rho g h V \xi_g$$

The volume energy density for a pumped hydro system will therefore depend on height h, which is given by

$$\frac{W_h = E_p}{V = \rho g h \xi_p^{-1}}$$

Assuming $\xi_p = 1$, the energy density for $h = 100\,\text{m}$ will be

$$W_h = 10^3\ \text{kgm}^{-3}\ 10\ \text{ms}^{-2} \times 100\ \text{m} = 10^6\ \text{J/m}^3$$

The upper limit v_{max} for the speed of the water as it enters the turbines can be calculated by equating the kinetic energy and the potential energy, assuming zero losses:

$$W_{\text{kin}} = W_{\text{pot}}$$
$$0.5\ m\ v_{\text{max}}^2 = W_{\text{pot}}$$

hence

$$v_{\text{max}} = \left(\frac{2W_{\text{pot}}}{\text{m}}\right)^{1/2} = (2gh)^{1/2} = 44 \cdot 3\ \text{ms}^{-1}$$

The pumped hydro storage turnaround efficiency is usually in the range of 70–85%.

Pumped hydro storage units were first installed by manufacturing industries in Italy and Switzerland in the 1890s to enable them to store surplus night-time output from run-of-river hydro power stations for use in meeting their peak power demand the following day. They were introduced into public electricity supply in a number of European countries during the early years of the last century, their role was to enable the economic despatch of thermal power plants in a power system.

6.2 The power extraction system

At first pumped hydro storage stations usually copied conventional hydroelectric design in having the power transformation (extraction) system located outdoors close to the lower reservoir. The increase in power capacity ratings and pumping heads, combined with the higher rotational speeds of turbines, has required the hydraulic unit to be set at considerable depths below the minimum tail water level in order to avoid cavitations. To meet these requirements, a massive concrete construction is needed to withstand the external water pressure and resist hydrostatic uplift, and therefore, an outdoor power transformation system has become increasingly expensive. Developments in civil engineering, particularly in the techniques of drilling, blasting, spill removal and rock support, as well as developments in SF_6 installed high-voltage switchgear and ducted busbars, make it cost-effective in many cases to erect the whole power transformation system – generating, transforming and switching plants – underground. This may involve a complex multi-cavern layout when plant dimensions are large and roof spans are limited by rock conditions. Moreover, this concept allows us to locate the power transformation system along waterways on which the dynamic response characteristics of the hydraulic system can be optimised in relation to construction costs. An outdoor station will still be necessary if the geology is unsuitable for underground construction. In that case, the power transformation system can be housed in one or more concrete-lined, vertical cylindrical shafts, excavated from the surface.

Surface pipelines, common in outdoor schemes, have been superseded by concrete-lined tunnels with the adoption of underground schemes. The only requirement is sufficient rock cover to provide the necessary additional resistance to internal pressure. Expenses for costly steel lining can be minimised, providing the shortest possible lengths of high-pressure tunnel consistent with overall constructional requirements. The choice between single- and multiple-supply conduits is an optimisation problem and depends on the economic balance of capital costs, hydraulic friction losses and plant availability. On the low-pressure side, where lower flow speeds are usually required, multiple tunnels are more favourable. Spinning reserve requirements dictate the necessity of rapid power loading and therefore short starting times (t_{st}) are needed for waterways. The starting time may be given by

$$t_{st} = \sum Lv/gh$$

where L = length of waterway section
v = flow velocity
h = head

It may be necessary to adopt relatively low water flow speeds in the sections of the hydraulic system not regulated by a surge chamber.

Before 1920, most pumped hydro storage systems were of the 4-unit type, in which the turbine-generator and pump-motor units were mounted on two separate shafts. This separate turbine-generator and pump-motor design is now seldom seen as its higher capital costs usually outweigh its advantages in respect of efficiency, availability and fast response.

After 1920, preference changed to favour 3-unit sets, which comprise a turbine, pump and generator-motor arranged together on a single horizontal or vertical shaft. Designs of this type have been widely adopted in Europe but are only used in a limited proportion of new installations where very high heads exist or where the shortest possible times for start-up from standstill and changeover between generation and pumping are required. During 70 years of exploitation the following few clear preferences have emerged. Horizontal-shaft sets provide easy access for maintenance, but for an outdoor station they involve expensive excavation since the pump unit must be deeply submerged. To avoid this, booster pumps have been employed to feed the main pumps in some cases although these are now not usually favoured due to additional operating complexity and reduced plant availability. Vertical-shaft sets, where the generator-motor is usually located above the turbine and the pump below, are now the most common arrangement for 3-unit sets in outdoor stations.

Francis turbines are available for heads up to about 700 m and further development will probably extend the head range to 900 m or more. Felton turbines are used for even higher heads. The highest single-head pumps are those operating under 635 m at Hornber, Germany. Multistage pumps can extend the head range to over 1 400 m.

A disengageable coupling of two different types is usually installed to enable the pump to be disconnected during generation. It may be a gear-type clutch as at Ffestiniog, Wales, operable only with the plant stationary. Alternatively, a hydraulic torque converter may be used to enable the pump to be brought up to speed and clutched in without stopping the turbine and generator-motor, as at Lunersee, Austria; when fast changeover is required it is normally rigidly coupled.

To start the machine the pumping mode is generally used, after which the load is transferred to the motor and the normally rigidly coupled turbine has to be dewatered to reduce rotational losses. At a number of stations, such as Hornberg in Germany, an overriding clutch enables the turbine to be disconnected and brought to standstill after the pump drive has been taken over by the motor. At the vertical-shaft Waldeck II pumped hydro plant, the location of the turbines above the generator-motor, with the pumps below, is a novel feature. Rigid coupling is used to achieve an exceptionally compact layout and the pumps are provided with retractable labyrinth rings to enable them to spin in air without cooling water during generation.

The first 2-unit set incorporating a reversible pump-turbine and motor-generator was introduced at Baldeney, Germany, in 1933. It was a small axial-flow unit. Only 21 years later a large reversible unit of 56 MW capacity operating under a head of 60 m was commissioned at Hiwassee, USA. Since then the single-stage pump-turbine has been the most widespread type of unit for heads up to about 600 m. The next step in the progress of the reversible pump-turbine was made in 1976 when the first 80 MW multi-stage pump-turbine operating under a head of 930 m was commissioned at La Coche, France. Subsequently, units of this type, rated at 150 MW operating under a head of 1 075 m, have been installed at Chiotas Piastra, Italy.

Compared with the 3-unit composition, the reversible 2-unit set allows a saving of about 30% on power plant capital costs by decreasing the number of hydraulic machines, main valves and penstock bifurcations. The drawback is that the pump starting regime is more complicated and the changeover time between generation and pumping is longer since the rotation of the turbine axis must be reversed. With the same rotational speed in both directions, the overall efficiency of the 2-unit set is lower than that of a 3-unit set; this is particularly noticeable with low-head plants when the range of head variation is high over the operating cycle. A two-speed generator-motor can provide the higher pump speed needed for optimal efficiency but requires higher capital cost and greater electrical complexity, which usually annuls the advantages of higher efficiency.

Reversible pump-turbines are not self-starting in the pumping mode and therefore complicated methods must be introduced. It should be mentioned that, in power systems, direct online starting is not usually feasible for generators with a rated capacity above 100 MW to avoid excessive system-voltage disturbance. An additional drawback is severe stressing of the generator windings. Reduced voltage starting by an autotransformer and series reactor could probably solve this problem but at the expense of additional complication, longer starting times and increased capital cost. The introduction of direct-connected Felton turbines or separate motor-generator sets for starting each main unit is an unnecessary expense.

Three methods are currently used for large hydraulic units:

- Pony-motor starting is widespread but increases rotational losses, despite a de-watered start: it requires more space and leads to a relatively long starting time.
- Back-to-back starting is a method used when the machines are run up synchronously in pairs, one as a pump and the other as a generator. The starting times are relatively short but the last unit has to be started by other means or not used for pumping at all (as at Piedilargo, Italy).
- A thyristor frequency converter is used to supply power at a progressively increasing frequency to each motor-generator, one after another until synchronous speed is reached. The pumping units are started de-watered and moderately fast starting times can be achieved. The same equipment can be used for dynamic braking. This method overcomes the drawbacks of the others and seems likely to become widespread.

Typical changeover times for large 3-unit sets with hydromechanical pump coupling and 2-unit sets with static thyristor frequency converter pump starting are shown in Table 6.1.

Table 6.1 Pumped hydro changeover times per set

Mode change	**Changeover times**	
	3-unit sets s	**2-unit sets s**
Standstill to full-load generation	120	120
Standstill to full-load pumping	180	600
Full-load generation to full-load pumping	120	900
Full-load pumping to full-load generation	120	480

Multi-stage reversible machines are less expensive than the alternative 3-unit arrangements and may be used for heads beyond the range of single-stage reversible sets, as at La Coche and Chiotas Pastra. However, they do have drawbacks. It is impossible to control generation in the discharge mode; guide-vane control in a pump-turbine with more than two stages is not cost-effective and therefore cannot be justified. Turbine efficiency will be substantially less than optimal if the fixed distributors are set for optimum pumping performance and pumped hydro generation will be correspondingly reduced. The absence of adjustable guide vanes also makes de-watered pump starting too difficult – watered back-to-back starting is the usual practice despite the fact that it requires a much longer starting time.

Fully regulated 2-stage reversible machines for heads up to 700–1 000 m have been developed to overcome these disadvantages. The first example is a 40 MW, 440 m head machine of this type for installation at Le Truel. A 300 MW machine has already been the subject of a preliminary design study by EdF (Electricité de France). A unique concept of unidirectional rotation is the Isogyre pump-turbine. The double-sided rotor consists of a turbine runner on one side and a pump impeller on the other. A single spiral casing is common to both operating modes and separate sleeve valves outside the runner and impeller are used as closing valves. The cost is lower than for a 3-unit set and mode changeover times are comparable. The largest machines installed to date are the 60 MW, 200 m head units in the upper stage of the Malta scheme, Austria.

Present and prospective future head limits of reversible pump-turbines are shown in Table 6.2.

Table 6.2 Head limits for reversible pump-turbines

Reversible pump-turbines	Present head limit (m)	Possible future head limit (m)
Single-stage, regulated	620	800–900
2-stage, regulated	440 (under construction)	1 000
Multi-stage, unregulated	1 075	1 500

Few high-head pump-turbines with rated capacity above 300 MW are yet in service and more experience is needed to understand whether continuing the increase of their parameters will entail significant loss of availability or increase in repair and maintenance costs. Progress in the increase of present limits of generating capacity and head for single-state reversible machines may continue to yield worthwhile capital cost savings, but at the same time, since the specific speed has to remain sufficiently high for optimum efficiencies to be achieved, serious problems of flow stability, vibration and component design and manufacture will be raised.

Usually the generator-motors used in the pumped hydro schemes have conventional air cooling, but since the unit capacities and heads for single-stage pump-turbines are increasing, water cooling may be needed. We are close to the limiting combinations of power capacity rating and rotational speed for which air cooling is feasible, and any further substantial increases will require the adoption of water cooling, which, incidentally, will enable us to raise the output limit for any given speed by about 60%.

6.3 The central store for pumped hydro

The choice of technically suitable pumped hydro storage sites and their optimum form of development is usually determined by topographical and geological characteristics, and may be constrained by a number of environmental limitations. The remoteness of available sites from the grid can also be crucial.

There are different methods of creating a reservoir. In a few cases, as at Kiev, Ukraine, Geestacht, Germany and Raccoon Mountain, USA, a large river has been used as the lower reservoir. The outfall has to be carefully designed to avoid any damage to navigation and water protection interests. Existing lakes, sometimes enlarged by dams across the outlets, are used in many schemes as both upper and lower reservoirs, as at Dinorwig in Wales. More commonly, a reservoir is formed by damming a river, due provision being made for the controlled release of floodwater and, where necessary, the passage of fish (as at Zagorsk, Russia or Luddington, USA). Soil or rockfill dams are now generally preferred to concrete structures because of appearance and, furthermore, their capital cost can be lower with modern construction techniques. Previous doubts about the stability of dams due to daily discharge and charge of the reservoir have been allayed by the development of impermeable upstream coverings and advances in the design of graded fill zones within the dam core. Asphalt or concrete multilayer linings with bottom and intermediate drainage layers and concrete-slab linings have been developed to a high level of reliability, and are widely used if the reservoir floor is not sufficiently watertight. Concrete-slab linings have been used in a number of schemes, such as Ronkhausen, Germany; the upper reservoir has been formed by levelling a hilltop and using the soil to form an enclosing dam. Such reservoirs are usually fully lined.

All the pumped hydro schemes developed so far successfully exploit suitably located reservoirs of water. Economic considerations usually dictate a choice of site for which the ratio of the horizontal to vertical distance between the upper and lower reservoirs is in the range of 4:1 but, in particularly favourable circumstances, it can be lower (as at Turlough Hill, Ireland, where this ratio is 2.5:1).

Consideration has been given to a number of alternative configurations that may be feasible for extending the range of siting possibilities, such as using the sea as a lower reservoir with an upper reservoir located on high coastal ground. Schemes of this type have been found to be more expensive than conventional alternatives, mainly because of increased costs related to corrosion protection and the prevention of seawater leakage from the upper reservoir to surrounding land, and unfavourable siting.

One of the new concepts being examined is underground pumped hydro storage, which would not be so limited by topographic and environmental constraints as conventional schemes. In the proposed concept (see Figure 6.2), a lower reservoir is located underground in hard rock, without any connection to a natural body of water. The upper reservoir may also be created artificially and requires less volume than a conventional one for the same energy capacity. The reason is that, since the energy capacity is directly proportional to the height of the water head, the distance between the lower and upper reservoirs is only constrained by head limits for pump-turbines and can be 1 000 m or more, compared with 300 m or less for typical surface-pumped hydro storage plants.

The cost of excavation would be significant, but might partly he offset by choosing a site close to a load centre, thereby saving transmission line length.

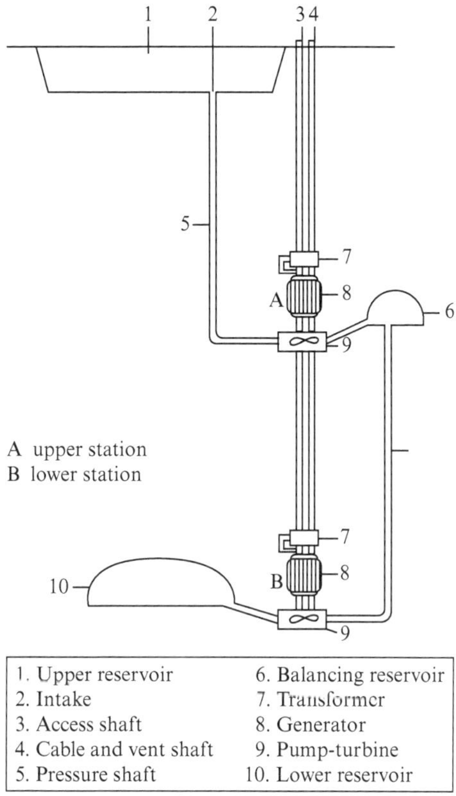

Figure 6.2 Schematic diagram of underground reservoir pumped-storage

The prospects for underground pumped hydro storage seem promising; practical cavern-excavation and tunnelling methods exist for the construction of the lower reservoir and high-head turbine technology is essentially developed (although further improvements would be desirable). The main geological constraint is the need to identify and predict the characteristics of suitable rock formations.

There is a theoretical possibility to merge the last two concepts in schemes that use the sea as an upper reservoir in conjunction with a lower reservoir underground. This concept may only be attractive if other siting opportunities have been exhausted and is definitely more expensive than the other alternatives.

6.4 Dinorwig

The 1 800 MW Dinorwig pumped hydro storage power station is one of the world's largest and has the fastest response of any pumped-storage scheme worldwide, being able to contribute 1 320 MW to the national grid within 10 s of demand. It is designed for a daily charge–discharge cycle, at a target efficiency of 78%. It takes about 6 h of full-load pumping to charge a full upper reservoir from empty, which, in turn, provides enough water for some 5 h of full-load generation. It is possible to start generation without any external power supply from the power system, so Dinorwig is capable of 'black start'.

The plant is designed to meet up to 40 operational mode changes per day, corresponding to 300 000 cycles over a 40 year fatigue life period. Dinorwig plant's six 330 MVA (PF, 0.95) vertical-shaft Francis reversible pump-turbine units operate at 500 rev/min. Its six 300 MW rated generator-motors operate at 18 kV and are connected in pairs, through 18 400 kV transformers and switchgear to the 400 kV busbar substation in the underground transformer hall. Each busbar is divided into three sections by SF_6 circuit breakers and each outer section is connected through a further circuit breaker to its own 400 kV underground cable, connecting Dinorwig electrically to the 400 kV outdoor substation at Pentir, about 10 km away.

The central store of Dinorwig pumped hydro comprises upper and lower reservoirs with their dams and tunnels. Its power transformation system consists of pump-turbines and generator-motors sited in underground caverns along with transformers, switchgear and pump starting equipment. Basic information about this station is given in Tables 6.3 and 6.4.

The fourth British pumped hydro storage station at Dinorwig was preceded by a station at Ffestiniog (4 × 90 MW), which was commissioned during the early 1960s. Both Dinorwig and Ffestiniog are located in North Wales and are connected, at 400 and 275 kV, respectively, to the UK National Grid 400/275 kV network. They are also connected through the Scottish 275 kV network with two more pumped-storage stations at Cruachan (4 × 100 MW) and Foyers (2 × 150 MW), both of which are owned by the North of Scotland Hydro-Electric Board.

In the 100 year history of pumped hydro storage exploitation, unit capacities have increased from a few tens of kW to over 400 MW, operating heads from less than 100 m to above 1 400 m and overall efficiencies from around 40% to well over 85%. Advances have also been made in the associated civil engineering techniques. The largest project currently being built has a generating capacity exceeding 2 000 MW.

There is over 90 GW of pumped storage in operation worldwide contributing about 3% of global electricity generating capacity. In the US pumped hydro accounts for about 2.5% of baseload generating capacity. There are some 150 operational pumped hydro plants of a total capacity of over 25 GW. It should also be mentioned that in 2010 about 5% of electrical energy consumed in the EU has passed through pumped hydro plants. Table 6.5 gives the main particulars of some notable projects completed or under construction before 2010. Data given in this table only prove the fact that pumped storage is a technology that 30 years after Dinorwig is still on the agenda. As to confirm this statement Scottish and Southern Energy (SSE) announced plans for two new pumped-storage developments to maintain a diversified portfolio of power stations, with the flexibility to respond to customer demand for electricity, while achieving a 50% reduction in the carbon

Table 6.3 Dinorwig's central store

Reservoirs	
Working volume	6 700 000 m^3
Maximum flow rate	420 $m^3 s^{-1}$
Maximum flow pumping	384 $m^3 s^{-1}$
Upper reservoir water-level range	33 m
Lower reservoir water-level range	14 m
Height of upper reservoir dam	36 m above existing ground level 68 m maximum height (toe excavation to crest)
Length of upper reservoir dam crest	600 m
Maximum width of upper reservoir dam at base	250 m
Height of lower reservoir embankment above existing ground level	3.5 m
Tunnels	
Length of low-pressure tunnel	1695 m
Diameter of low-pressure tunnel	10.5 m
Depth from top of surge pond structure to invert of high-pressure tunnel	558 m
Dimension of surge pond top	80 × 43.324 m
Dimension of surge pond base	73.25 × 38.5 m
Depth of surge pond structure	14 m
Diameter of surge shaft	30 m
Depth of surge shaft	65 m
Diameter of orifice shaft	10 m
Depth of orifice shaft (invert of surge shaft to low-pressure tunnel)	40 m
Dimensions of orifice	10 × 45 m
Diameter of high-pressure shaft	10 m
Depth of high-pressure shaft (invert of low-pressure tunnel to invert of high-pressure tunnel)	439 m
Diameter of high-pressure tunnel	9.5 m
Diameter of high-pressure penstocks	3.8 m at manifold 2.5 m at main inlet valve
Average length of high-pressure system (high-pressure shaft to pump-turbine)	700 m
Number of draft tube tunnels	6
Average length of draft tube tunnels (pump-turbine to bifurcation)	164 m
Diameter of draft tube tunnels	375 m downstream of the pump-turbine 5.8 m at the bifurcation
Average length of tailrace tunnels (bifurcation to tailworks structure)	382 m
Number of tailrace tunnels	3
Diameter of tailrace tunnels	8.25 m
Diameter of ventilation shaft	5 m
Depth of ventilation shaft	255 m
Excavations	
Main underground excavation	106 m^3 (approx. 3 × 10^6 tons)
Central tip excavation	1.71 × 10^6 m^3 (approx. 3.5 × 10^6 tons)
Wellington flat excavation	950 000 m^3 (approx. 2 × 10^6 tons)
Haford Owen flat excavation	950 000 m^3 (approx. 2 × 10^6 tons)
Other lake margins	610 000 m^3 (approx. 1.25 × 10^6 tons)
Infill to lake bed	1.65 × 10^6 m^3 (approx. 3.3 × 10^6 tons)

Table 6.4 Dinorwig's power transformation system

Underground caverns	length	71.25 m
Machine hall	width	23.5 m
	height	59.7 m (max.)
Transformer hall	length	161 m
	width	23.5 m
Number of interconnecting busbar galleries	3	
	length	46 m
	width	14 m
	height	17.5 m
Main inlet valve gallery	length	146 m
	width	6.5 m
	height	18.5 m
Number of starting equipment and	2	
heating and ventilating galleries	length	36 m
	width	14 m
	height	26 m
	length	28 m
	width	14 m
	height	29.5 m
Draft tube valve gallery (between the SE and	length	172.7 m
heating and ventilating galleries)	width	8.5 m
	height	23.6 m
Pump-turbines		
Type	Reversible Francis pump-turbine	
Number	6	
Plant orientation	Vertical spindle	
Average pump power input	28.3 MW	
Pumping period	6.31 h	
Synchronous speed	500 rev/min	
Generator-motors		
Type	Air-cooled	
Excitation	Thyristor rectifiers	
Sequence control	REED relay or solid-state type	
Generator-motor switchgear		
Type	Air-blast	
Fault breaking capacity	120 kVA	
Current rating	11.5 kA	
Voltage	18 kV	
Average generated station	1 681 MW	
Generating period at constant output (all machines operating)	5 h	
Station power requirements when generating	28 MW	
Standby operational mode	Spinning in air	
Energy load pick-up rate	0–1.32 MW in 10 s	
Minimum submergence below lake level	60 m	
Generator-motor transformer		
Number	6	
Approximate rating	340 MVA	
Voltage ratio	18 kV/400 kV	
Transmission switchgear		
Type	SF_6 metal clad	
Breaking capacity	35 000 MVA	

(*Continues*)

Table 6.4 (Continued)

Current rating	4 000 A
Voltage	420 kV
Pump starting equipment	
Type	Static variable frequency converter
Rating	11 MV (in addition, back-to-back starting is provided for emergency use)
Auxiliary system voltages	11 and 3.3 kV and 415 V

Table 6.5 Pumped hydro projects in service or under construction up to 2010

Station	Country	Pumping head (m)	Unit number capacity type (MW)	Station capacity (MW)	Year of commission
Imaichi	Japan	52.4	3 × 350-VR(1)	1 050	1984
Kiev	Ukraine	66		225	1966
Ludington	USA	98	6 × 312-VR(1)	1 872	1974
Zagorsk	Russia	100	6 × 200-VR(1)	1 200	1988
Foyers	UK	182	2 × 150-VR(1)	300	1974
Taum Sauk	USA	240	2 × 230-VR(1)	460	1963
Juktan	Sweden	260	1 × 334-VR(1)	334	1978
Vianden	Luxembourg	287	9 × 105-HF/C(2)	1 141	1959
Racoon	USA	317	4 × 383-VR(1)	1 532	1978
Mountain Ffestiniog	UK	320	4 × 90-VR(1)	360	1963
Cruachan	UK	360	4 × 100-VR(1)	400	1966
Waldeeck II	Germany	329	2 × 220-VF/C(1)	440	1974
Rodund II	Austria	324	1 × 283-HF/C(2)	283	1976
Bath Country	USA	387	6 × 457-VR(1)	2 740	1985
Robiei	Switzerland	410	4 × 41-VR(1)	164	1968
Drakensbergs	South Africa	473	4 × 270-VR(1)	1 080	1981
Helms	USA	495	3 × 358-VR(1)	1 070	1981
Numaparra	Japan	508	3 × 230-VR(1)	690	1973
Ohira	Japan	512	2 × 256-VR(1)	510	1975
Okuy Shino	Japan	539	6 × 200-VR(1)	1 200	1978
Dinorvig	UK	545	6 × 300-VR(1)	1 800	1982
Tamaharo	Japan	559	4 × 300-VR(1)	1 200	1983
Hondawa	Japan	577	2 × 306-VR(1)	612	1983
Bajina Basta	Yugoslavia	621	2 × 315-VR(1)	630	1983
Hornberg	Germany	635	4 × 248-HF/C(1)	992	1974
La Coche	France	931	4 × 80-VR(5)	320	1976
Chiotas	Italy	1 070	8 × 150-VR(1)	1 200	1980
Piastra Edolo	Italy	1 260	8 × 127-VR(5)	1 016	1981
San Fiorano	Italy	1 404	2 × 125-VP/C(6)	250	1974
Bailianha	China	192	4 × 306	1224	2006
Guangzhou	China		8 × 300	2400	1993
Shisanling	China		4 × 200	800	1995
Tianhuangping	China		6 × 306	1836	2001
Tongbai	China		4 × 300	1200	2006
Xiangshuijian	China	363	4 × 250	1000	2011

V: vertical shaft; H: horizontal shaft; F: Francis turbine; C: centrifugal pump; R: reversible pump-turbine; (): number of stages dioxide intensity of electricity produced. SSE said it was aiming to submit planning applications for the two developments in 2011 and would therefore seek guidance from the Scottish government on proposals to build a 300 and 600 MW pumped-storage plant in the Great Glen. Both projects would require new dams to be built. To limit the visual impact of the development the pumping and electricity generation infrastructure would be built underground.

Chapter 7
Compressed air energy storage

7.1 General considerations

Citywide compressed air energy systems have been built since 1870. Cities such as Paris, Birmingham, Offenbach, Dresden in Germany and Buenos Aires in Argentina installed such systems. Victor Popp constructed the first systems to power clocks by sending a pulse of air every minute to change the pointer. They quickly evolved to deliver power to homes and industry. As of 1896, the Paris system had 2.2 MW of generation distributed at 550 kPa in 50 km of air pipes for motors in light and heavy industry. Usage was measured in metres. The systems were the main source of house-delivered energy in these days and also powered the machines of dentists, seamstresses, printing facilities and bakeries. The application of elastic energy storage in the form of compressed air storage for feeding gas turbines has long been proposed for power utilities; a compressed air storage system with an underground air storage cavern was patented by Stal Laval in 1949. Since that time, only two commercial plants have been commissioned; Huntorf CAES, Germany, and McIntosh CAES, Alabama, USA.

To return the electrical energy stored in an intermediate form to an electrical power system, an energy-conversion process from this form is needed.

In this case, compressed gas is a medium that allows us to use mechanical energy storage. When a piston is used to compress a gas, the energy is stored in it. This energy can be released when necessary to perform useful work by reversing the movement of the piston. Pressurised gas therefore acts as an energy storage medium.

The ideal gas law relates the pressure P, the volume V and the temperature T of a gas as follows:

$$PV = nRT$$

where n = number of moles
R = gas content.

According to the first law of thermodynamics, which may be given by

$$Q = \delta U + A$$

change in internal energy δU *is* equal to the sum of the heat Q and the useful work A. Therefore, the amount of useful energy that can be stored or extracted from the

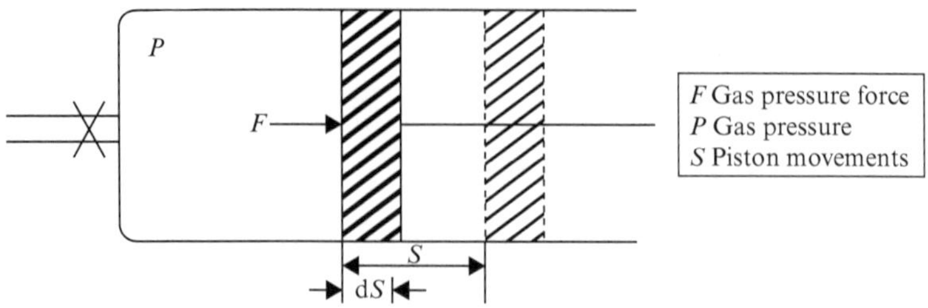

Figure 7.1 Pressurised gas moves a piston by applying a force

compressed gas depends on the type of process applied since some heat exchange is always associated with any gas process. The only exception is an adiabatic process, which does not really occur since it would require infinitely large insulation.

Let us assume that the piston can move as shown in Figure 7.1 without friction. It is then possible to derive the work done when the piston is forced to move a distance S, compressing the gas. The definition of the pressure P from force F acting on the piston with area A may be written as

$$P = \frac{F}{A}$$

hence

$$W = \int F ds = \int PA ds = \int P dV$$

This integral would be easy to solve if P were constant during the process; that is we would essentially have an isobaric process where the work done by the gas was

$$W = P \int_{V_1}^{V_2} dV = P(V_2 - V_1)$$

Here $V_2 - V_1$ is the volume traversed by the moving piston. Movement of the piston in the opposite direction, against the gas, applying an external force, leads to the storage of a corresponding amount of energy (see Figure 7.2*a*).

If the process follows an isotherm instead of an isobar, the gas law transforms to Boyle–Mariotte's law (see Figure 7.2*b*):

$$PV = nRT = \text{constant}$$

hence

$$W = \int_{V_1}^{V_2} P dV = nRT \int_{V_1}^{V_2} \frac{dV}{V} = nRT \ln\left(\frac{V_2}{V_1}\right)$$

Let us calculate the order of magnitude of the volumetric energy density for a compressed gas cylinder with a starting volume V_0 of 1 m^3 and a pressure P_0 of 2.03×10^5 Pa. If the gas is compressed to a volume of 0.4 m^3 at constant temperature, the amount of stored energy is

$$\frac{W}{V_0} = \left(\frac{1}{V_0}\right) nRT \int_{V_0}^{V} \frac{dV}{V} = \left(\frac{1}{V_0}\right) P_0 V_0 \ln\left(\frac{V_0}{V}\right) = P_0 \ln\left(\frac{V_0}{V}\right) = 1.86 \times 10^5 \text{ J}$$

This energy density is much higher than those of magnetic or electric fields, as we shall see later.

It is difficult to maintain an exactly constant pressure or temperature; therefore, an arbitrary reversible process mentioned in Figure 7.2 is most interesting from the practical point of view. The energy can be determined by the area under the curve, irrespective of the slope, between the two volumes V_1 and V_2 and the corresponding temperatures T_1 and T_2.

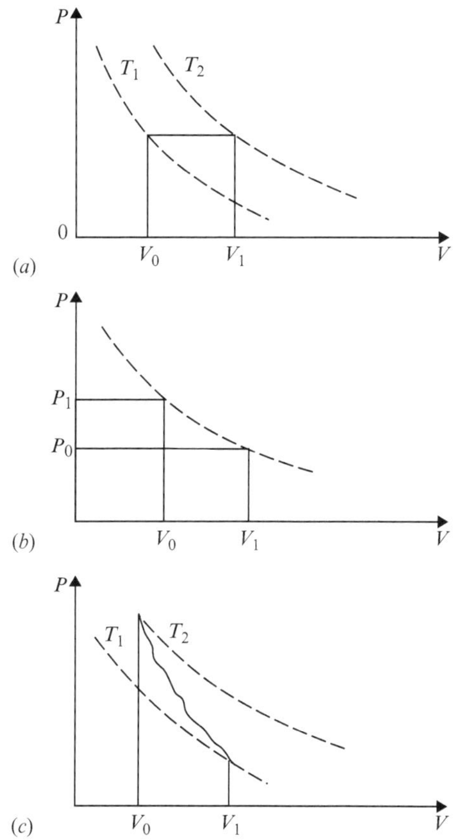

(*a*) Isobaric process $\delta P = 0$; $W = P(V_1 - V_s) = nR(T_h - T_l)$
(*b*) Isothermal process $\delta T = 0$; $\delta T = 0$; $W = nRT \ln(V_1/V_s) = nRT \ln(P_1/P_s)$
(*c*) Arbitrary real slope (reversible) process W = the area under the curve = energy stored in the gas

Figure 7.2 Types of ideal gas process (n moles of gas are involved)

The compressed air energy storage (CAES) concept involves a thermodynamic process in which the major energy flows are of work and heat, with virtually no energy stored in the compressed air itself. The performance of a CAES plant depends on the precise details of both the compression process and the expansion process.

7.2 Basic principles

To demonstrate the main features of the CAES concept, let us consider a gas turbine power plant that basically comprises four components, as shown in Figure 7.3*a*, namely a compressor, a combustion chamber, a turbine and an electricity generator. The compressor is used to deliver atmospheric air at high pressure to the combustion chamber. Fuel is injected and burnt in the same combustion chamber, heating high-pressure air. Hot high-pressure gaseous products from the combustion chamber drive a turbine from which they are exhausted at roughly atmospheric pressure. Approximately two-thirds of the mechanical energy

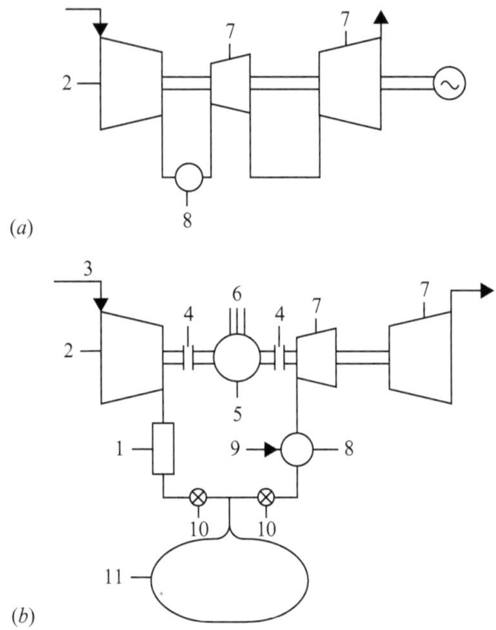

(*a*) Simple-cycle gas turbine
(*b*) Simple-cycle gas turbine modified to CAES configuration

1. Cooler	7. Turbine
2. Compressor	8. Combustor
3. Air	9. Fuel
4. Clutch	10. Valve
5. Generator/motor	11. Air storage cavity
6. Power supply	

Figure 7.3 Development of the CAES concept

generated by the turbine is used to drive the compressor. The remaining one-third is converted by an electricity generator to electricity.

The same components exist in a CAES arrangement as in a gas turbine, but some new elements, namely clutches and pressure vessels as shown in Figure 7.3*b,* are added. Clutches are added so that the compressor and turbine may be separately connected to the generator, which also has to work as a motor. Pressure vessels are needed to store compressed air.

In this way, compression and expansion of the flow medium (air) take place at separate times. This separation has the advantage that all of the output of the gas turbine is available as useful energy.

Large-scale energy storage in the form of compressed air, stored in natural or artificial underground caverns, is economically and technically viable at present. The energy stored in the compressed air can be utilised using low-pressure turbines, but this approach results in very low turnaround efficiencies. The alternative is to expand the air through a combustor.

The system functions in the following manner: a compressor that draws its energy from the power system during the demand trough at night, for instance, takes in atmospheric air, compresses it to a high pressure and passes it to the central store – a pressure vessel. At times of peak demand the air is taken from the container, heated in a combustor by burning fuel and then expanded in the turbine to the ambient pressure. The mechanical work thus produced is converted into electricity in a generator driven by the turbine, and fed to the power system.

The effect of heating the air in the combustor by burning fuel is that the useful work is between 20% and 40% of the work expended in pumping.

In a CAES scheme all the energy produced by the turbine is available for electricity generation at times of peak demand, since the compressor has done its work during a demand trough. This requires a larger generator-motor installed capacity compared with the set of generator, compressor and turbine in a conventional gas turbine arrangement.

The volume of the air reservoir is determined by the amount of energy to be stored due to the power system requirements. The rated capacity of the compressor can also be varied according to the required length of time during which it charges the reservoir again. The volume of the reservoir might, for example, be sized to run the turbine for 1 h at full load, while the compressor is designed to refill the container in 4 h, so that the compressor is sized for only one quarter of the turbine's air throughput and the charging ratio of 1:4 is implemented. The air storage gas turbine can be designed for various charging ratios (1:1, 1:2, 1:4, etc.) according to the utility's operational needs and the geological conditions of the site.

In addition to this flexibility of design, the main consequence of change to the conventional turbine's composition is that, on a specific cost per rated power capacity basis, the main part of the equipment costs about one-third of an equivalent gas turbine, since the electrical output of CAES is three times greater. The potential economic attractiveness of CAES depends mostly on being able to take advantage of this low equipment cost despite the cost of air containment, clutches, heat exchangers and other equipment.

7.3 The central store

First developed for storing natural gas and various noxious wastes, the technology of 'solution mining' of salt cavities to a reasonably controlled shape is well established and provides a very cheap method of excavation for large storage volumes. There is now considerable experience of salt cavities being used to store gas, oil and other substances in Europe and the United States. The influence of cavern size, depth and operating pressure, as well as salt properties, on the basic cavern design is well established.

Salt caverns are practically leak-tight: it is estimated that the leakage losses at Huntorf CAES are between 10^{-5} and 10^{-6} per day. The use of salt caverns as central stores is particularly attractive; however, the creation and exploitation of salt cavities has several potential problems – namely brine disposal, cavern rat-holes, creep and turbine contamination.

Solution-mining technology, based on fresh water dissolving salt and becoming saturated with it, involves drilling a bore into a salt cavern, cementing the upper side to the surrounding rock, jetting water down a central bore and then removing brine, as shown schematically in Figure 7.4. The brine has to be disposed of

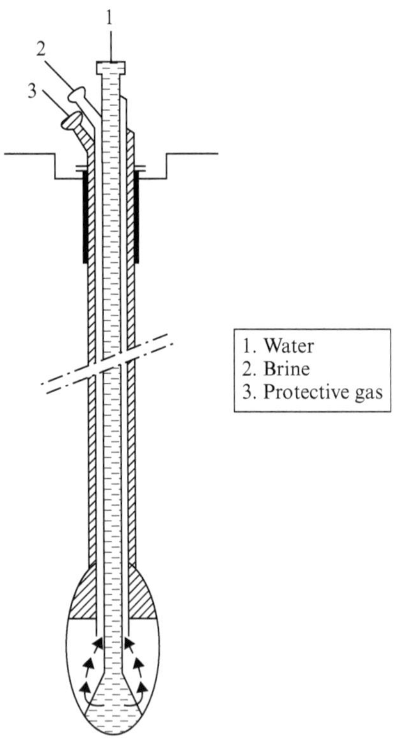

Figure 7.4 Schematic diagram of leaching process for an air reservoir in an underground salt deposit

somewhere on an environmentally acceptable site; the chemical industry is the preferable solution.

The developing cavern is monitored by sonar, and when a desirable diameter has been achieved, a cover gas that is of lower specific weight than the brine (and does not react with salt) has to be pumped in to make the solutioning proceed downwards. The ideal shape of the cavern should be a vertical cylinder with an aspect ratio of 6:1. It should be mentioned that certain variations in solubility are encountered in the form of layers of anhydrite or the presence of high-solubility potassium- or magnesium salt. Layers of anhydrite may remain as platelets protruding into the developing cavern, whilst the presence of high-solubility potassium- or magnesium salt may cause 'ratholes' – in the worst case even extending the planned cavern surface to a neighbouring cavern. Such problems can be avoided by drilling test bores and analysing the brine during solutioning, which of course increases the overall capital cost.

Creep or ceiling collapse could also be reasons for premature closure of a CAES central store. The risk of salt carry-over can also create problems in turbine design. Experience at Huntorf and McIntosh, after years of operation at 1000 cycles per annum shows that no problems of either turbine contamination due to salt carry-over or creep of the walls have been noticed, which is very encouraging for future CAES prospects.

Salt layers at the right depth, of the right thickness and suitably located where storage plant is needed are quite common in European and North American geology, but this is not so worldwide, e.g. Japan. Other possibilities do exist, however, for the creation of underground caverns. Reasonable low-cost technology can be used in aquifer regions where there is suitable domed caprock, which is well known for natural gas storage.

Depending on the permeability of the porous rock medium, a number of wells have to be drilled into the aquifer region to develop an air bubble that will displace the water contained. This process is rather slow, but after the creation of an air bubble, since air has a much lower viscosity than water, it is possible to achieve the required charge and discharge rates necessary for CAES work in a power system. It is known that to keep total pressure losses over a charge–discharge cycle to around 10–20 bar, no less than 50 injection wells may be required, which, together with surface connections, will be the most expensive part of the essentially low-cost aquifer air storage concept, shown in Figure 7.5.

A third possible approach to underground air storage is the use of mined cavities in hard rock. This is rather expensive since it requires mining activity and spoil disposal from underground on a very large scale. Nevertheless, the high cost is offset by the possibility of operating at 'constant' pressure. In this case varying the free volume is required, which can be achieved by using a water-compensation leg. It is desirable to have reasonable quality rock to avoid water in-leakage becoming a problem. A schematic arrangement is shown in Figure 7.6. The environmental effects of water supply and varying the water level in the reservoirs at ground level have to be taken into account.

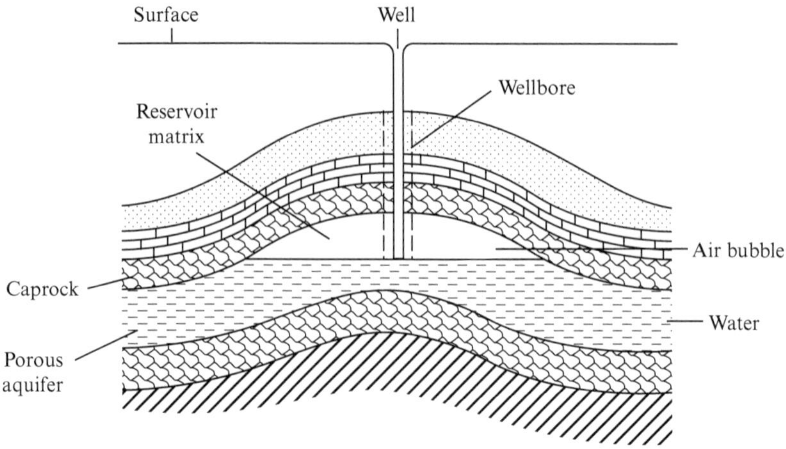

Figure 7.5 Schematic diagram of aquifer air storage

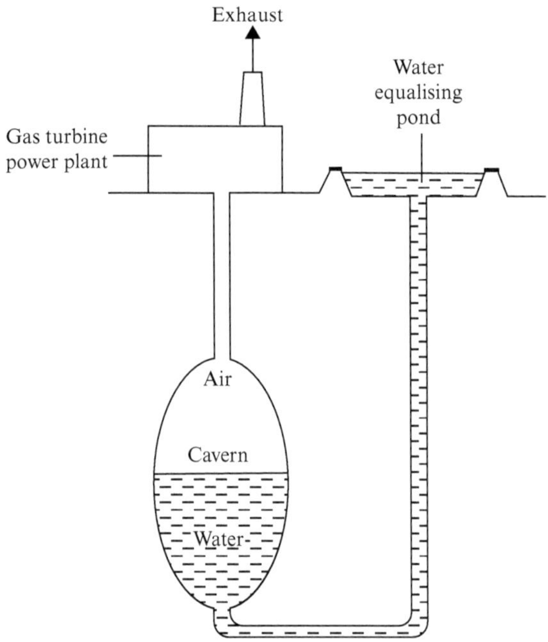

Figure 7.6 Air storage gas turbine power plant with a constant-pressure reservoir

A potential hazard, as yet not fully quantified, is the so-called champagne effect. During the discharge period, air in a partially filled cavern will be dissolved in the compensating water. After a certain time the water becomes saturated and, as it is displaced up the compensation leg during the next charge period, bubbles will be released. This is potentially hazardous, although it has been shown that it can be resolved by proper design and good management of the storage plant operating cycle.

Unfortunately, leak-tight rock is difficult to find so there will usually be leakage losses. The use of a 'water curtain', as shown in Figure 7.6, is a promising approach capable of reducing leakage considerably. Despite requiring remedies such as the water curtain, cavern lining or others, the overall cost of hard-rock air storage will probably not be much higher than for salt cavities or aquifers. The problem is that there is no practical experience in the design and operation of such types of CAES.

The CAES facilities at Huntdorf and McIntosh use engineered salt caverns as the storage reservoir for the compressed air, although other types of geological formation are thought to be suitable, including sandstone aquifers and depleted natural gas fields. EPRI studies show that more than half of the United States has geological conditions that are potentially suitable for CAES.

While the use of salt domes has been proven at both existing CAES plants, the engineering involved in preparing the cavern can take up to two years. In Ohio, Norton Energy Storage, a firm backed by private equity group Haddington Ventures, is proposing to use a limestone mine for a 2 700 MW CAES development. Although tests carried out in 2 000 on the mine's rock formations indicate that it would be well suited to air storage, the project is still under development.

A special requirement for CAES siting is to locate cavities geologically suitable for air storage. Environmental concerns, such as visual impact and noise, and technical problems like cooling-water supplies and closeness to transmission lines and fuel supplies, are issues for any generating plant in a power system, and will be no more severe than for conventional gas turbine installations which are already designed to very high standards in each of these respects.

Deep water in lakes and the ocean can provide pressure without requiring high-pressure vessels or drilling into salt caverns or aquifers. The air goes into inexpensive, flexible containers such as plastic bags below in deep lakes or off sea coasts with steep drop-offs. Obstacles include the limited number of suitable locations and the need for high-pressure pipelines between the surface and the containers. Since the containers would be very inexpensive, the need for great pressure (and great depth) may not be as important. A key benefit of systems built on this concept is that charge and discharge pressures are constant and depend on depth. The University of Nottingham carries a research work on seabed–anchored containers funded by the one of the Europe's leading power and gas companies E.O.N.

CAES is potentially a useful, efficient and low specific capital cost storage means, but widespread future application depends on opportunities for siting and the relative economic benefits connected to the potential for development to satisfy the growth of fuel costs and availability criteria. A number of studies carried out in the United States have shown that there are no technical limitations to the widespread application of CAES, but since large volumes of air have to be stored, conventional vessels are quite impractical and underground air storage in salt, hard rock or other 'impermeable' strata is essential. These methods are expensive and so the required volume has therefore to be minimised by choosing the highest practical storage pressure. Current turbine and compressor design limitations and underground cavity costs make storage pressures of 70–80 bar likely to be the most cost-effective.

7.4 The power extraction system

There have been a number of major studies of CAES schemes, all designed to displace gas turbines from peaking and low-merit generation duties: typically 1 500–2 000 h operation and often less. With differences in detail, owing to different machinery assumption or siting considerations, all these studies have come to much the same conclusion on the principal design features.

First, it should be noted that CAES is neither a pure storage, for example pumped hydro, nor a pure generating plant like a conventional gas turbine. Rather, it is a combination of pure storage and conventional gas turbine. As a pure storage device it requires a central store, discussed above, and a power transformation system in the form of gas turbine and compressor both connected through clutches and possibly gearboxes to a motor-generator.

The compressors and turbines used are not simple modifications of conventional gas turbines but rugged industrial machines with two or more casings in both the compressor and the turbine train. In the case of compressors, unless impractically large air flows and unit ratings are postulated, it is also necessary to incorporate a gearbox to run the high-pressure machines at well above synchronous (3 000 rev/min) speed. Without this, the compressor stage efficiency will drop dramatically, but the gearbox also imposes a size limitation.

These features are all present in the first CAES plant in the world, operated by the West Germany utility Nordwestdeutschen Kraftewerke (NWK) at Huntorf near Bremen, and in the second operated by the Power South (formerly Alabama Electric or AEC) Cooperative near McIntosh.

Advanced CAES concepts are subject to the same design constraints as in plants that already exist, namely high storage pressures and the use of as much existing machinery as possible. The main components are discussed below.

The highest pressure ratio currently available for uncooled industrial compressors is 17:1. In this case air is delivered at a temperature of 430–450 °C. This can be achieved with two axial machines in series, operating from an atmospheric inlet.

The high-pressure compressor would comprise radial machines, equivalent to those currently available, which would need to be geared to higher than synchronous speed. Current experience limits the compressor's power to 70 MW for a delivery pressure about 75 bar because of gearbox constraints.

It is possible to take advantage of developments in industrial gas turbine technology towards obtaining a higher pressure ratio (up to 25:1) for the synchronous speed low-pressure and intermediate-pressure compressors. In this way, the cycle becomes 'more' adiabatic, which leads to a decrease in fuel heat rate of up to about 200 kJ/kWh. There is still the possibility that the HP compressors could be high-speed axial machines driven via gearboxes operating without special cooling, using a pure adiabatic cycle instead. As well as the compressors, the turbines would need to be slightly different from those already in existence.

During the charge period, the PTS is formed by an electric motor that drives a compressor. During the discharge period, the PTS is a turbine that drives an electricity generator.

During the charge regime, the storage medium – air – leaves the compressor substantially heated – the heat generated during compression is rejected by intercoolers and cool air is stored in the cavern. When the plant is generating electricity during the discharge period, the air must be heated up by burning natural gas.

Several intercoolers therefore have to be employed in the compressor train. This enables approximately isothermal compression, and minimises the charge energy requirements to within practical limits. As is known, the specific compression work W_c is directly related to the temperature rise across the compressor and may be given by

$$W_c = C_p \delta T_c = C_p T_{in}(P_r^{(\gamma-1)/(\xi,\gamma)} - 1)$$

where T_{in} = absolute inlet temperature
δT_c = air temperature increase due to compression
P_r = pressure ratio
$\gamma = \dfrac{C_p}{C_r}$
ξ_c = polytropic compression efficiency.

Without intercooling each successive stage will have increasing inlet temperature. The air temperature will accordingly be raised, leading to a rise of specific input work W_c. By reducing T_{in} before each stage it is possible to decrease the overall specific compression work W_c. For example, without intercooling, 70:1 compression from 15 °C would lead to a δT of about 810 °C and require a work input of 870 kJ/kg. With three intercoolers (as at Huntorf), no stage has an outlet temperature above 230 °C and the compression work is reduced to 550 kJ/kg; a saving of 37% is thus achieved. To obtain this saving and prevent damage to the central store the air requires cooling.

The intercooler required in the compression cycle rejects heated air energy as low-grade heat. However, the cooling-water requirements are considerable. If a 300 MW CAES unit of current design and equal charge/discharge times were installed on one site, the cooling water needed would be equivalent to a conventional thermal power unit of about 125 MW. This would require either a suitable coastal site or the use of cooling towers, which could become the dominant sight of the CAES. If the desired cooling-water quantities are not available, it would be possible to introduce an air-cooling system, but then considerable attention would have to be given to the noise produced.

Owing to the need to cool compressed air during the charge regime, the specific fuel consumption per mass of air during the discharge regime is higher than in a conventional gas turbine. Nevertheless, since the net output from the CAES turbine is three times that of a conventional one, the overall fuel consumption is significantly decreased.

A parameter such as storage turnaround efficiency is practically meaningless for assessing the performance of a CAES. Instead, two measures of CAES performance have to be introduced.

Let us define the charge energy factor f_{cef} as the ratio of generated electricity output E_{d} to electric energy used for charging E_{c}:

$$f_{\text{cef}} = \frac{E_{\text{d}}}{E_{\text{c}}}$$

For a pure storage unit this factor is exactly the same as turnaround efficiency. It should be mentioned that, for CAES, it is more than unity while for other storage methods it is less than unity.

According to the definition, the fuel heat rate f_{fhr} is a ratio between thermal energy of fuel consumed E_{f} in the CAES and generated electricity output E_{d}:

$$f_{\text{fhr}} = \frac{E_{\text{f}}}{E_{\text{d}}}$$

Fuel heat rate is a common characteristic for thermal energy plant (gas turbine included) and is

$$f_{\text{fhr}} = \frac{3600}{\xi_{\text{cp}}}\ \text{kJ/kWh}$$

where ξ_{cp} is the thermal efficiency of conventional plant.

The overall fuel cost f_{cd} of energy generated by CAES is given as follows:

$$\frac{f_{\text{cd}} = f_{\text{cc}}}{f_{\text{cef}} + f_{\text{c}} f_{\text{fhr}}}$$

where f_{cc} is the charge energy cost and f_{c} is the cost of fuel used by CAES during a discharge regime.

CAES, with a high f_{cef} where fuel savings are minimal, is the best peak power source if the basic mid-merit and some base-load generation are provided by coal-fired plants; that is nuclear power does not become the dominant energy source. CAES plant burns no fuel at all and can achieve a similar generating cost, but its capital cost is considerably higher. However, in the economically preferable 'high-nuclear' scenario, low f_{fhr} is preferable to high f_{cef}.

There are two main approaches to CAES development that need to be considered, regardless of the fuel to be used. First, increases in turbine firing temperature in conventional CAES plant will lead to significant increases in the charge energy factor. Second, we can effectively conserve charge energy by thermal storage and so reduce or ideally eliminate fuel consumption. Which is preferred will depend on technical developments, fuel cost and future fuel policy. Let us consider the potential of each in general terms and determine which development is more promising.

Substantial improvements in both charge energy factor and fuel heat rate are possible with very little change to the conventional gas turbine design. As a working body for a high-pressure (HP) turbine expanded air should have a temperature of 550 °C. Before reaching the low-pressure (LP) turbine it will be reheated to 825–900 °C. These parameters are currently available for turbines for CAES, based on experience at Huntorf. There is only limited industrial experience of higher temperature HP turbines, but LP turbines are now in use at 1 100 °C. Since the specific output from a turbine is proportional to the absolute inlet temperature, higher firing temperatures would increase the ratio of generated energy per mass of stored air, and hence improve the charge energy factor (CEF). For example, by increasing the LP turbine firing temperature to a still modest 900 °C, the CEF can be raised to 1.3.

The next step in the development of CAES technology must be to reduce the quantity of high-quality fuel needed for reheating the air in the expansion phase. One way to improve efficiency is to transfer thermal energy from the power-generating turbines' exhaust gas stream and use it to preheat the air expanding to the HP combustor. Such a 'recuperator' feature is included in the current design for the McIntosh CAES and should reduce fuel consumption by 25%. This leads to a substantial reduction in fuel heat rate from 5 800 kJ/kWh at Huntorf to less than 4300 kJ/kWh at McIntosh.

For Huntorf-type CAES turbines, any additional fuel consumption will be used to increase the generation at a very high efficiency, but this also leads to a small reduction in fuel heat rate. The influence on CEF and fuel heat rate of increasing the low-pressure turbine inlet temperature, and both LP and HP inlet temperatures, is shown in Figure 7.7, for variable pressure air storage schemes. It can be concluded from Figure 7.7 that, with this technology, the charge energy factor could be raised to 1.7 with a firing temperature of 1 100 °C, although that would presumably involve substantial changes to gas turbine design. Such an f_{cef} level would make exceptionally efficient use of limited low-cost charge energy. However, it is most unlikely that this approach will result in better than 5% reduction in fuel consumption compared with the recuperated compressed air storage concept, which is currently used in McIntosh CAES design.

Further development can either involve the substitution of synthetic fuels for oil or a change in the basic cycle by recycling the heat released when the air is compressed and which is currently exhausted to the atmosphere. This ideally adiabatic cycle, using thermal energy storage (TES), forms a logical development towards eliminating fuel from the CAES operation. According to this so-called 'near-term hybrid' concept the HP expander would not be fired at all and would receive preheated air only from the exhaust recuperator. The inlet temperature would typically be 320 °C and the air will exhaust from the HP turbine at about 130 °C. The casing of the HP turbine would have to be arranged so that the HP

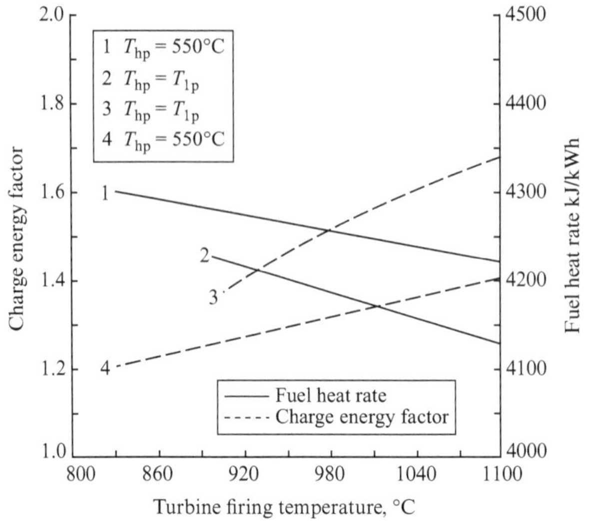

Figure 7.7 Effect of increasing LP and HP temperatures on the charge energy factor

exhaust could pass out to the TES, from which air preheated to about 420 °C could be returned to the LP combustor. This combustor, LP turbine and recuperator would be exactly the same as in currently available 'state-of-the-art' CAES plants. An important facility not included in conventional CAESs would be an additional control/shut-off between the TES and the LP combustor because the large volume of the TES would lead to intolerable turbine overspeed levels in the case of a generator's sharp load decrease.

The key to achieving significant reduction in fuel consumption is to conserve more effectively the energy taken from the grid during the charge regime. To achieve approximately isothermal compression and a minimal specific charge energy factor, intercooling is required. Without intercooling, when compression is practically adiabatic, the specific work input during the charge regime is significantly increased. The compressor delivery temperature would be 850–900 °C for an 80:1 pressure ratio. If air were taken directly from the compressor and expanded in a turbine, which means no storage at all, up to 80% of the charge energy would be recovered without any additional fuel.

Some companies like RWE, together with GE, are developing the advanced adiabatic CAES technology (AA-CAES), which captures heat from the compressed air to make the cycle more efficient. In AA-CAES concept the heat rejection is set aside by using minimal intercooling during the compression phase and capturing the heat in a TES system. The compressed air flows over a solid material – such as ceramic bricks – so that the heat can be transferred, and when the air is discharged from the cavern it flows back over the material to preheat it before it passes to the expander.

To create a storage plant, it is necessary to store the air in such a way that both its pressure and temperature are maintained; TES is therefore needed. In that case, during the charge regime the TES works as an intercooler, cooling the compressed air to achieve acceptable air temperatures of around 30 °C for low-cost underground storage caverns and, during the discharge regime, it reheats the air to the required temperature for the HP turbine inlet.

Fairly large TES plants have been used successfully in applications as diverse as air heating for blast stoves; however, the problems of scale, in addition to technical constraints for high-pressure and high-temperature turbines and compressors, mean that pure adiabatic cycles are not currently available. More realistic cycles would involve lower pressure ratios and some form of supplementary heating to achieve adequate turbine inlet temperatures. There are two mechanisms to achieve this. The first concept, which uses some fuel to heat the air after the TES, the so-called conventional and adiabatic hybrid, is shown in Figure 7.8*a*. The second alternative involves the use of additional charge energy to heat the TES

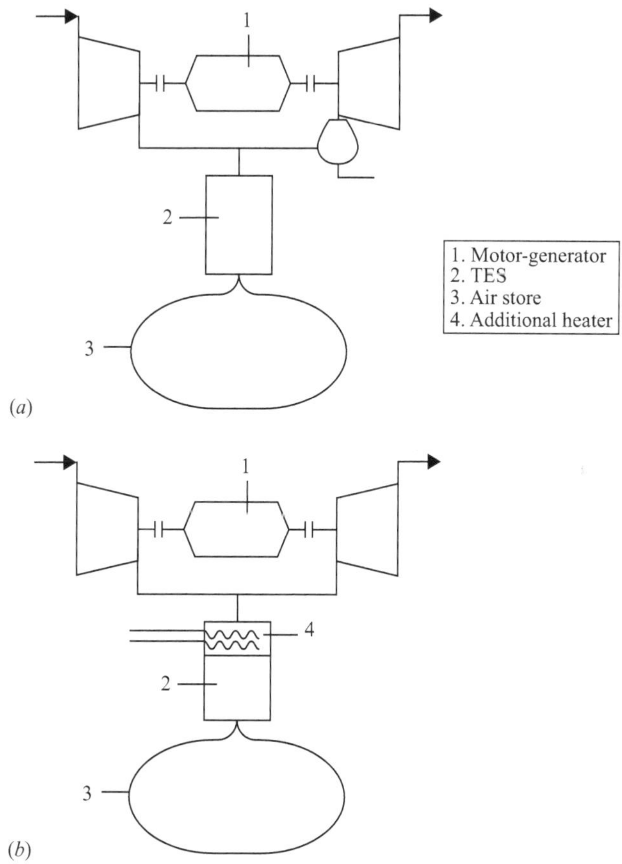

(*a*) Adiabatic hybrid CAES
(*b*) 'Heat-topped' CAES

Figure 7.8 CAES combined with TES

directly to beyond the compressor delivery temperature, so-called heat topping, and is shown in Figure 7.8*b*. The stored air, a ready source of low entropy, enables either heat topping energy or fuel consumed by the adiabatic hybrid to be recovered at between 50% and 65% efficiency according to the pressure ratio.

There are a number of proposed TES concepts for CAES usage. In the cheapest one, only a heat exchanger to a secondary storage medium is pressurised, and the central store of the TES itself is at atmospheric pressure. This is the so-called indirect type of CAES. Direct contact schemes involve cycling air, which passes directly through a packed bed of high-density, high specific heat capacity material, so that the central store is placed in a number of pressure vessels. The high-pressure central store for the direct contact TES is quite expensive, but owing to the very high cost of materials for the heat exchanger and the loss of efficiency with the indirect types, direct contact TESs are preferable.

It is crucial to the overall efficiency of an advanced CAES cycle that the TES has a turnaround effectiveness of no less than 0.9. This requires materials of reasonably good conductivity with high surface area/volume ratios and with good air-to-solid heat-transfer coefficients. 'Checker bricks', conventionally used in hot blast stoves, are designed to a low pressure-drop specification and therefore do not achieve adequate performance for CAES. According to EPRI information, 'pebbles', manufactured from a dense fireclay or cast iron, of a modest size would probably lead to the most efficient use of TES material and the smallest pressure-vessel volumes. According to estimates, a TES unit for a 300 MW CAES plant with 8.64×10^{13} J energy capacity would require a containment of 17 m diameter and 17 m high, made of concrete (like a nuclear reactor pressure vessel), containing some 22 000 tons of matrix. Such a TES would achieve an efficiency of no less than 0.93 and a turnaround temperature loss of less than 20 °C.

An alternative concept closer to current practice would use externally insulated low alloy steel pressure vessels – 12 vessels of 5 m diameter and 17 m length. This is only four times the volume of the alumina pebble-bed TES that NASA have been operating for 15 years at their establishment on Moffat Field Air Base in California.

Development of AA-CAES is not without its challenges, however. The main challenge is the compressor outlet temperature, which reaches around 620 °C. Although the European study suggested development of a completely new machine, the feasibility study has investigated the possibility of developing the compressor section using mainly standard components. But the future is for the second-generation CAES. They will use standard components – including the compressor and gas turbine – and will be totally scalable so that plant size will be flexible according to geology.

7.5 Two industrial examples

7.5.1 Huntorf

The Huntorf CAES (HCAES) is shown schematically in Figure 7.9 and was engineered by Brown Boveri Co., Mannheim, Germany, as a 'minimum risk' commercial prototype.

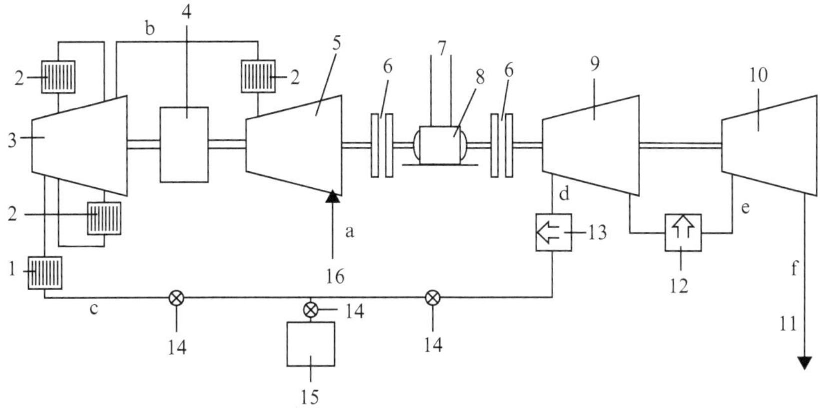

1. Aftercooler
2. Intercooler
3. Compressor (high-pressure stage)
4. Gearbox
5. Compressor (low-pressure stage)
6. Clutch
7. Transmission line
8. Motor/generator
9. High-pressure turbine
10. Low-pressure turbine
11. Exhaust
12. Low-pressure combustor chamber
13. High-pressure combustor chamber
14. Valve
15. Air cavity
16. Intake

Power conditions
a. 15 °C, 1 bar
b. 55 bar
c. 37 °C, 68 bar
d. 550 °C, 43 bar
e. 825 °C, 11 bar
f. 390 °C, 11 bar

Figure 7.9 Huntorf CAES

Construction at Huntorf was started in May 1975 and the equipment was installed during the period July 1976 to September 1977, and, following some commissioning delays necessitating some equipment rebuilding, the plant was finally handed over to NWK in December 1978. On a future installation, no more than three years should be necessary for the complete construction and commissioning programme. The plant has a design specification of $f_{cef} = 1.2$ and $f_{fhr} =$ 5 800 kJ/kWh. The whole project cost was considerably less than the equivalent cost in gas turbines in Germany at the same price date.

The air reservoirs at Huntorf can be operated only as sliding-pressure containers with pressure in the reservoir sliding from 66 to 46 bar, and having a pressure gradient of 10 bar/h when the turbine is running during discharge. During the 2 h when the turbine is intended to run at full load, the pressure drop is 20 bar.

As the reservoirs are recharged by the compressors, the pressure in them rises so that the compressors' charge pressure also increases from 46 to 66 bar during the 8-h charging time.

The air pressure in a sliding-pressure reservoir changes during the CAES charge and discharge modes. Therefore, to maintain a constant electrical output, the turbine is run at a constant inlet pressure of 46 bar. The pressure in the reservoir is therefore throttled to the turbine inlet pressure, and allowance is made for the resulting throttling loss. Despite the throttling loss and the changing condition of the air in the reservoir during charging and recharging, the result of heating the air in the two combustors, as mentioned earlier, is that the useful work is up to 20% greater than the work done during charging (when a constant-pressure reservoir is used this value can rise to 40%).

For a generator output of 290 MW and a turbine inlet pressure of 46 bar, assuming the pressure in the reservoir varies by 20 bar, a reservoir volume of about 130 000 m^3/h is needed.

The top of the salt deposit at Huntorf lies at a depth of about 500 m. Owing to the geological requirements of a 100 m thick layer above the reservoirs, they are placed at a depth of 600 m. Two vertical cylindrical shaped reservoirs, each with a capacity of some 135 000 m^3, were leached out of the salt, using the brine method described. Each reservoir has a diameter of about 30 m and a height of approximately 200 m.

The horizontal distance between the power station and the reservoir is 200 m, with 180 m separating the reservoirs themselves. The two reservoirs are linked to the power station by pipes and are normally operated in parallel.

Owing to the underground (petrostatic) pressure acting on the reservoirs and the plastic state of the salt, the volume of the cavities will probably decrease, halving the original volume each 4 000 years.

The volume of the air reservoirs, and hence their installation cost, is governed not only by the required energy capacity of the plant but also by the pressure and temperature of the air in the reservoirs. It has been proved that a storage pressure of between 40 and 60 bar, where the ratio of work output to work input reaches its most favourable value, is economically optimal. Above this range of pressure the required fuel consumption would in fact decrease, but the cost of the machinery required would be considerably greater. The considerable increase in the large volume of air storage required makes its cost too high for pressures below 400 bar.

HCAES is designed to generate peak power for around 2 h per day, and to be charged over 8 h at a reduced mass flow and compressor power rating. This means that the compressors are designed for only a quarter of the turbine throughput, and so the charging ratio is 1:4.

The pressure range from 40 to 60 bar is much higher than the inlet pressure of conventional gas turbines, which operate at a pressure of about 11 bar. Therefore, the largest available conventional gas turbine, an axial machine running at synchronous speed of 3 000 rev/min, and its combustor, may be used only as the low-pressure turbine and low-pressure combustor to utilise the pressure drop from 11 to 1 bar.

The pressure drop from 46 to 11 bar had to be handled by an HP turbine but no examples of gas turbines were available in 1976. Consequently, the high-pressure turbine is based on a small intermediate-pressure steam turbine design with 550 °C inlet temperature. The high-pressure combustor inlet has a pressure of 46 bar.

At the Huntorf CAES the air from the reservoirs is heated in the HP and LP combustors by natural gas. Exhaust gases from the LP turbine are not used for preheating air from the reservoir. Reducing stations bring gas to the required pressure before entering the combustors. Light fuel oil is also suitable for installations of this kind.

The compressor group comprises an axial low-pressure unit and a centrifugal HP machine driven via a 45 MW gearbox at 7 626 rev/min. The total drive power is 60 MW. The compression process incorporates intercoolers. For geological reasons the air is cooled further to 50 °C in an aftercooler before passage to the reservoirs.

The generator has an apparent power of 341 MWA and a voltage of 21 kV. It is separated from both turbine and compressor by self-engaging clutches, so that the generator may be used as a motor to drive the compressors. Detailed information about Huntorf CAES equipment is given in Table 7.1.

If the inlet valves are opened, the air coming from the reservoir will be burned in the HP combustor, thus accelerating the turbine. The LP combustor may be fired and the turbine loaded only after reaching synchronous speed and synchronising the generator.

The whole starting procedure from standstill to full load normally takes 11 min, or about 6 min with a quick start. Synchronous speed is reached in 2 min in both cases.

The compressors reach synchronous speed in about 6 min when starting from standstill. They are started with the aid of the turbine using the remaining air in the reservoir, which accelerates the turbine-driven generator to synchronous speed. When the generator is synchronised it can be run as a motor driving the compressor. The turbine has then to be shut down and the clutch between turbine and generator opens.

The Huntorf plant is fully automated and remotely controlled from the load-dispatching station in Hamburg. No personnel are required at the power station.

The method of controlling an air storage gas turbine is quite different from that of a conventional gas turbine. In the latter case the output is regulated by varying the turbine's inlet temperature, while the air throughput remains almost constant. The result is a relatively high heat consumption at part loads. In contrast, in an air storage gas turbine the air flow rate is matched to the required output, and the inlet temperature to the HP turbine is held constant together with the exhaust temperature of the LP turbine. Part-load heat consumption is then much better, so the air storage gas turbine can also be used efficiently to regulate power.

7.5.2 McIntosh

The first completely new, commercially viable generation option was delivered to the US power utilities when the multistage, reversible McIntosh CAES system was commissioned on 27 September 1991 by the Alabama Electric Cooperative (AEC), after 2 years and 9 months of construction.

Table 7.1 Principal data for the Huntorf CAES machine set and air container

Gas turbine	
Type	Single shaft with reheat
Capacity	290 MW
Speed	3 000 rev/min
Air throughput	417 kg/s
Inlet conditions, HP turbine	46 bar/550 °C
Inlet conditions, LP turbine	11 bar/825 °C
Exhaust temperature after LP turbine	400 °C
Specific heat consumption	1 400 kcal/kWh
Fuel	Natural gas
Generator	
Apparent power	341 MVA
Power factor	0.85
Voltage	21 kV
Speed	3 000 rev/min
Cooling	Hydrogen
Compressors	
LP compressor	
Type	Axial
Speed	3 000 rev/min
Intake conditions	10 °C/1.013 bar
Air throughput	108 kg/s
HP compressor	
Type	Centrifugal
Speed	7 626 rev/min
Conditions after compression	48–66 bar/50 °C
Number of intercoolers	3
Number of aftercoolers	1
Drive power of both compressors	60 MW
Air reservoir	
Number of reservoirs	2
Height of reservoir	200 m
Reservoir diameter	30 m
Volume of reservoir	135 000 m^3
Depth of reservoir	600 m
Distance between reservoirs	180 m
Horizontal distance between reservoirs and power station	200 m

The McIntosh CAES (MICAES) was intended to improve upon the technical specification of the HCAES design as it contains a recuperator (not installed in the HCAES). The recuperator uses waste heat to preheat air from the cavern before it is heated in the combustor chamber, reducing this premium fuel consumption by about 25%. The remaining basic components of the MICAES, including its compressor, combustor and expansion turbine, are similar to those found in HCAES. It has a rated capacity of 110 MW and can change load at a rate of 33 MW (30%) per minute, which, on a percentage basis, is three times faster than any other type of power plant. MICAES is designed to operate efficiently when partly loaded; for example, it loses only 15% in efficiency when loaded at 20% of this rated capacity. A conventional coal-fired power plant, in comparison, would lose about 50% of its efficiency at such partial load. One full charge from the 110 MW MICAES plant can provide enough electricity to meet the demand of 11 000 homes in the southern state for 26 h. But, perhaps more importantly, it provides its owner, the Power South (formerly AFC), with a useful tool to manage the resources of its system in an effective and economic manner.

Air storage for MICAES, located 500 m below ground in a massive subterranean salt layer that is 8 miles deep and about 1.5 miles in diameter, was created by solution mining. The volume of this vertical cylindrical shaped cavern, 300 m tall and 80 m in diameter, is more than 5.32×10^6 m^3. Air pressure in the cavern varies from 74 to 45 bars and compressed air flows through the turbine at a rate of 170 kg/s.

7.6 Despatch and economic limitations

The time required to bring the Huntorf CAES to full load in the discharge regime is around 6 min, which is less than that required for the equivalent industrial-type gas turbine but more than that achievable by peaking gas turbines based on airplane turbines. The rapid start time at Huntorf is achieved by increasing the air flow from the cavern beyond the nominal design value until the combustor reaches its nominal temperature. In this case full load can be achieved earlier than the designed firing temperatures. This leads to a slight reduction in turnaround CAES efficiency but allows quite a modest increase in firing temperature, thus reducing blade and rotor stresses and therefore increasing reliability and lifetime. The ability to increase generation rapidly, provided it can also be done reliably, makes CAES plant more attractive than conventional gas turbines. Despite the fact that CAES is not as effective for rapid loading – pumped hydro storage can be designed for about 10 s response and flywheels can be even faster – the Huntorf plant, for example, is regularly started up to five times per day to provide the necessary assistance in power despatch in its 'home' power system.

Possibly the major drawback of compressed air storage is that it requires clear fuel such as natural gas or distillate oil. These 'premium' fuels are likely to become increasingly expensive. Their availability for power generation at present varies from country to country and, in the long term, they are likely to become less attractive for electricity generation. However, it should be mentioned that the quantities required for this generation are not as great as might be supposed; it has been estimated that CAES could smooth the load on the British system if up to 5% of the total fuel burnt was high-grade fuel to power CAES plant. The performance and value of CAES can only be compared with other storage technologies on a comparison of (fuel plus charge) energy cost and capital cost.

Neglecting, for the moment, the possible shortage of oil and gas, it is interesting to compare the fuel costs for CAES plant and widely used pumped storage.

The first notable point is that a CAES scheme such as Huntorf has a significantly higher fuel cost than pumped hydro if distillate oil or natural gas prices are still higher than the price of coal. This does not mean that the overall economics of pumped hydro plants are better, since one must take into account capital and operating costs, as well as factors such as the limited numbers of sites that may be available.

CAES plants are less sensitive than pumped storage schemes to whether the nuclear element of the system has grown such that nuclear power is available for charging storage schemes at night.

Two-shift coal-fired plants are likely to have lower fuel costs than CAES plant of the Huntorf type, but use of adiabatic CAES plant may become competitive with pumped hydro plants.

It should be feasible to heat the expanding air not with gas or oil but with coal or derived synthetic fuels. Coal substitution can be achieved in a number of ways. Fluidised-bed combustion, promoted for gas turbine and combined-cycle applications, could be used instead of oil combustion chambers. Separate gasification or coal liquefaction could be employed to produce a substitute fuel requiring no modification to the CAES turbine. In each case, the capital cost of the conversion equipment is high; low-cost coal is needed to make the substitution economically attractive. Alternatively, the promising concept of water electrolysis may be considered for the generation of synthetic fuels. The best economic proposition arises when a completely separate synthetic fuel production plant is employed, together with fuel storage, so that it is loaded optimally even though the CAES plant may demand only peaking fuel supplies. Such an approach can be applied to any type of CAES and favours cycles with the smallest demand for fuel. Synthetic fuels will be considered in detail in Chapter 8. Only three CAES plants have been built to date, of which two have been briefly described here. The third, a 25 MW demonstration CAES at Sesta, Italy, was built in porous rock. At present it is out of action. A Japanese utility planned to build a 35 MW 6 h commercial CAES unit using a porous rock medium for storage; it was expected to be commissioned in April 1997. There are additional CAES plants built or planned. Research has been done in Israel to build a 300 MW CAES facility using hard-rock aquifers. Similar projects have been started elsewhere to look into the possibilities of CAES systems. Another plant currently under development is being designed by Norton Energy Storage LLC in the USA. Their site is a 10 000 000 m limestone mine 700 m deep, in which they intend to compress air up to 100 bar before combusting it with natural gas. The first phase is expected to be between 200 and 480 MW and cost \$50 to \$480 million. Four more stages are planned, to develop the site to a possible capacity of 2 500 MW. A project that appears to be advancing is the Iowa Stored Energy Park, which is being developed by a group of municipal utilities with the backing of DOE funds through Sandia National Laboratories. This project will be the first CAES plant to use an aquifer as the storage reservoir. The cavern is a sandstone aquifer with a tight shale caprock.

There are also provisional plans to build a 1 050 MW CAES in Russia.

It may be concluded that CAES plants will have a limited, but valuable, role in future power systems.

Chapter 8

Hydrogen and other synthetic fuels

8.1 General considerations

Synthetic fuels are considered to be substitutes for natural gas or oil and are made from biomass, waste, coal or water. Production of these fuels demands energy, which can be obtained from base-load power plants during off-peak hours. Therefore, synthetic fuels are a type of energy storage since it is possible to use them instead of oil or gas for peak energy generation. The fuels themselves are only a type of medium (hydrogen, e.g. is simply a method to store and transmit energy); as with any other storage concept, a power transformation system and central store are also required. Storage media have to be produced during off-peak hours in a chemical reactor or electrolyser – this has to be considered as a part of a power transformation system used during the charge regime.

During the discharge regime, the storage media have to be converted into electrical energy, using any kind of thermal plant with an appropriate combustion chamber. As mentioned in Chapter 7, CAES is among the possible consumers of synthetic fuels.

These storage media have to be stored in a special containing device – a central store. The use of synthetic fuels does impose some problems of safety and container material, but is not very different from the infrastructure of storage and distribution systems involving natural gas and oil fuels.

8.2 Synthetic storage media

Among synthetic fuels, the most promising are methanol (CH_3OH), ethanol (C_2H_5OH), methane (CH_4), hydrogen (H_2), ammonia (NH_3) and methylcyclohexane.

Production of methanol from coal requires conversion of the coal into carbon oxides and hydrogen:

$$\begin{aligned} C + H_2O &\rightarrow CO + H_2 \\ CO + 2H_2 &\rightarrow CH_3OH \\ CO + H_2O &\rightarrow CO_2 + H_2 \\ CO_2 + 3H_2 &\rightarrow CH_3OH + H_2O \end{aligned}$$

The mixtures of CO, CO_2 and H_2 can be reacted in the so-called Lurgi process under 50 atm pressure and 250 °C temperature, or in Cu/Zn catalysts fluidified in liquid hydrocarbon.

Ethanol can be produced from CO_2 and H_2 by upgrading the hydrogen content in the same way as for methanol.

Biomass is an alternative basic raw material attractive in regions devoid of coal. Methanol can also be produced from biomass:

$$C_xH_yO_z + 2\left(\frac{x-z}{2}\right) H_2O \leftrightarrow CO_2 + \left[\frac{y}{2} + 2\left(\frac{x-z}{2}\right)\right] H_2$$

The process involves destructive distillation of organic solids in the presence of water, and is catalysed finally by cobalt molybdate.

Ethanol can also be produced from biomass by fermenting simple sugars to alcohol and carbon dioxide (CO_2). The preliminary hydrolysis process to convert cellulose into monosaccharides involves either a combination of heat and acid or enzyme fermentation. The latter method has been chosen for a government research programme in Brazil and has led to a 20% alcohol substitution in transport applications.

One of the most promising synthetic fuel and energy storage media for the foreseeable future is hydrogen. It can be derived from waste using any source of low-quality energy and can be used as a fuel in a very environmentally friendly way; it can be combusted back to water in a simple and clean reaction without any pollution.

The main drawback is the extreme flammability of hydrogen and the problem of storing the gas under pressure. Liquefaction could simplify storage but no less than 30% of charged energy would thereby be consumed. Hydrogen is flammable within a wide range of concentrations in air (4–75 vol%) whereas, for example, the range for methane is 5–15 vol%. Hydrogen also has a very low ignition energy (approximately 0.07 times that of methane).

A brief comparison of synthetic fuels is given in Table 8.1.

8.3 Hydrogen production

Hydrogen can be produced by several industrial methods:

- catalytic steam reforming of natural gas;
- chemical reduction of coal:

$$C + H_2O \rightarrow CO + H_2$$

or

$$CO + H_2O \rightarrow CO_2 + H_2$$

- industrial photosynthesis;
- ultraviolet radiation;
- partial oxidation of heavy oils;
- thermal decomposition of water with the help of thermochemical cycles; and
- electrolytic decomposition of water.

The last of these methods comprises two processes:

cathode reaction:

$$2H_2O + 2e^- \rightarrow H_2 + 2OH^-$$

Table 8.1 Comparison of synthetic storage media

Fuel type	**Liquid density (kg/m³)**	**Energy density**		**Boiling point (°C)**	**Mass transport efficiency (%)**	**Comments**
		10^7 J/kg	**10^9 J/m³**			
Methanol	797	2.1	15.8	64	25	Flammable gas, bad smell, corrosive, could be used as a fuel
Ethanol	790	2.77	21.0	79	30	Flammable gas
Methane		3.6	15.12	–164	25	Flammable gas, could be used as a fuel or energy input to release H_2
Ammonia	771	1.85	14.4	–33	17.6	Toxic gas, bad smell, regeneration of H_2 requires input of energy
Hydrogen gas at 150 atm, 20 °C	–	14	1.7	–252	100	Explosive gas
liquid at	–252 °C					
	71	14	10.5	–252	100	Flammable liquid
Metal hydride	–	Up to 1.1	up to 0.021	–	–	H_2 regeneration requires energy input

anode reaction:

$$2OH^- \rightarrow \frac{1}{2}\,O_2 + H_2O + 2e^-$$

According to Faraday's law of electrolysis, the mass of hydrogen m discharged may be calculated:

$$m = \frac{1}{F} \times \frac{A}{Z} \times I \times t$$

where F = 96 500 Cb/kg equivalent (Faraday's constant)
A = atomic weight
Z = valency
I = electric current through the electrolyte
t = duration of electrolysis.

The overall reaction enthalpy involving the transfer of two electrons per molecule of hydrogen (atom of oxygen)

$$H_2O \rightarrow H_2 + \frac{1}{2}O_2 - 237\ \text{kJ/kmol}$$

determines the theoretical minimum cell voltage: at 1 bar, 25 °C, $G = 237$ kJ/mol (2.016 g of H_2), requires 2 Faraday of electrons (193 kCb); this charge must be passed through 1.23 V, which is the equilibrium voltage of the cell. At higher temperatures the cell voltage will be lower since energy will have already been supplied as heat; for example, at 1200 K the equilibrium voltage is approximately 0.9 V. In practice, the equilibrium voltage has to be exceeded if any positive gas evolution is to occur. Typically, the voltage is in the range 1.5–2.05 V.

The amount of hydrogen that could be produced by a 60 kW power transformation system is 25 g or 280 l/min of gaseous H_2 plus half that amount of oxygen (in volume terms) at atmospheric pressure, neglecting thermal losses due to the current flow through the electrolyte. The electric current required is 40 kA and the cell voltage is 1.5 V.

Despite the fact that the predominant methods used now are catalytic steam reforming of natural gas (the 'commercial-grade' gaseous hydrogen, which has carbon- and sulphur-containing impurities) and partial oxidation of heavy oils, electrolytic and thermal decomposition of water are more suitable for the use of hydrogen for energy storage. In this case, hydrogen will be produced from water using the energy generated by either large hydro plants or base-load nuclear- or coal-fired power plants. It is even possible to use renewable photovoltaic-, solar- or wind-generated energy for hydrogen production (see Table 8.2 for hydrogen production methods comparison).

After transmission to the consumer, hydrogen will be used as a primary fuel for peak energy generation or simply as an alternative to other fuels. Combustion as a fuel would result in the recombination of hydrogen and oxygen to form water, thus completing the cycle. This hydrogen-based power utility concept (hydrogen economy) is shown schematically in Figure 8.1 and is under consideration in a number of scientific institutions. One of the main advantages of a hydrogen economy is that hydrogen can replace the fossil fuel burned in internal combustion engines and turbines, thus converting chemical energy into kinetic or electrical energy. Hydrogen can be used in conventional internal combustion engines, thus replacing conventional fuels. In case of fuel cells usage, the latter being electrochemical devices have a theoretical efficiency advantage over heat engines. Fuel cells production is more expensive compared with internal combustion engines but may become cheaper as new technologies will be developed. Fuel cells require high-purity ('technical-grade') hydrogen because the impurities would degrade the fuel cell life time. Currently (as for 2010 technologies) it takes 2.5 times as much energy to make a hydrogen fuel cell than may be obtained from it during its life time.

Hydrogen can be used on-site for greenhouse-neutral gaseous fuels direct (non-biological) production. Thus, captive-hydrogen-mediated production of greenhouse-neutral methane has been proposed (note that this is the reverse of the present method of acquiring hydrogen from natural methane, but one that does not require ultimate burning and release of fossil fuel carbon). Green plants, in a sense, already use solar power to make captively produced hydrogen, which is then used to make easier-to-store-and-use fuels. In the plant leaf, solar energy is used to split water into hydrogen and oxygen, the latter gas being released. The hydrogen

Table 8.2 Hydrogen production methods

	Natural gas	Nuclear	Solar	Wind	Biomass	Coal
Primary source	15.9 million cu. ft. of natural gas – only a fraction of current US annual consumption	240 000 tons of unconcentrated uranium, five times today's global production	2 500 kWh of sun per square metre per year, found in the Southwestern states of the Sun Belt	7 m/s average wind speed, typically found in many parts of the country	1.5 billion tons of dry biomass (initially by-products such as peanut shells, then concentrated crops)	1 billion tons of coal – which would require doubling current US domestic production
Infrastructure	777 000 fuelling stations; though a more likely scenario would include a mix of larger central production plants	2 000 600 MW next-generation nuclear power plants; only 103 nuclear power plants operate in the States today	113 million 40 kW systems, covering 50% of more than 300 million acres – an area three size the size of Nevada	1 million 2 MW wind turbines, covering 5% of 120 million acres, or an area larger than California	3 300 gasification plants, and up to 113.4 million acres – or 11% of US farmland – dedicated to growing the biomass	1 000 275 MW plants; only 12 sites were proposed for a DOE demonstration plant – not all met the requirements
Total 2010 cost	$1 trillion	$840 billion	$22 trillion	$3 trillion	$565 billion	$500 billion
Price per GGE (gallon of gas equivalent)	$3.00	$2.50	$9.50	$3.00	$1.90	$1
CO_2 emissions (measured in tons)	300 million	0	0	0	600 million*	600 million†
2010 status	There are four fuelling stations that now produce hydrogen from natural gas	The first very high temperature Reactor in the United States will be built at Idaho National Laboratory in 2021	Honda built an experimental solar-powered hydrogen-refuelling station at its laboratory in California in 2001	A 100 kW turbine is now being built at the National Renewable Energy Laboratory in Colorado	Government funded biomass research will be transferred to private industry in 2015	By 2012, the first FutureGen demonstration plant should be running at 50% capacity

*Zero net emissions because crops pull CO_2 from the air.

†90% will be captured and stored underground.

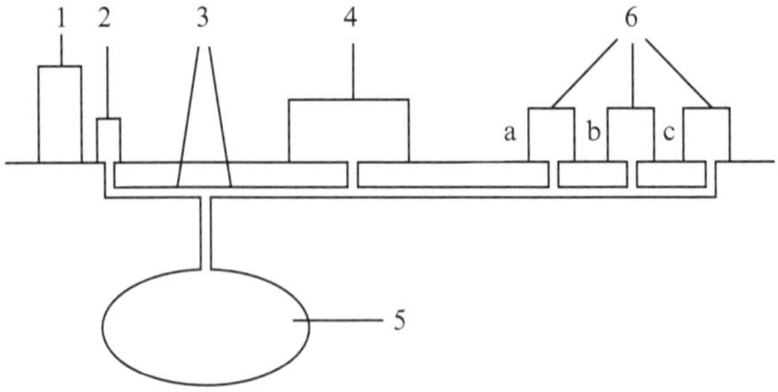

1. Base power source
2. Electrolyser
3. Underground transmission (tanker shipment as an alternative)
4. Liquid H_2 storage
5. Underground gaseous H_2 storage
6. H_2 consumers:
   a. Small conventional power plants
   b. Synthetic fuel industry
   c. Domestic fuel

Figure 8.1 Hydrogen-based power utility concept

produced is then used ‘on-site’ by the plant to reduce CO_2 from the air into various fuels, such as the cellulose in wood, and the seed oils that are the basis for vegetable oil, biodiesel, etc. Captive hydrogen (and CO_2) may be used on-site to *synthesise* methane, using the Sabatier reaction. This process is approximately 80% efficient, reducing the round trip efficiency to approximately 20–30%, depending on the method of fuel utilisation. This is even lower than hydrogen, but the storage costs drop by at least a factor of 3 because of methane’s higher boiling point and higher energy density. Liquid methane has 3.2 times the energy density of liquid hydrogen and is easier to store. Additionally, the pipe infrastructure (natural gas pipelines) are already in place. Natural gas-powered vehicles already exist, and are known to be easier to adapt from existing internal engine technology than internal combustion autos running directly on hydrogen. Experience with natural gas-powered vehicles shows that methane storage is inexpensive, once one has accepted the cost of conversion to store the fuel. However, the cost of alcohol storage is even lower, so this technology would need to produce methane at a considerable savings with regard to alcohol production.

Methane may be formed from the reaction of hydrogen with carbon or its oxides, and can be prepared from hydrogen by the Fisher–Tropsch process. This exothermic reaction releases 205 kJ of heat and can be written as follows:

$$CO + 3H_2 = CH_4 + H_2O$$

Under certain conditions, CO_2 may react in a similar way to form combustible hydrocarbon fuel.

Methane may also be synthesised from hydrogen and carbon by the following exothermic reaction, with 73 kJ heat release:

$$C + 2H_2 = CH_4 \qquad \delta H = -73\ \text{kJ}$$

Methane could probably be used as an energy source since it is already a major fuel with an established distribution network, but would require condensation to liquid for efficient transportation if produced at a location remote from an existing natural gas distribution network.

Methane may be used by consumers as an ordinary fuel, releasing 890 kJ of heat, or enriched with hydrogen by up to 5% to prepare a clean fuel with relatively low CO_2 emission. It does not have such high flammability as hydrogen. The reaction is as follows:

$$CH_4 + 2O_2 = CO_2 + 2H_2O$$

If it is essential that there be no emission of oxides of carbon in the production of a methane-based fuel, the following methods may be considered:

- decomposing methane to hydrogen and carbon with 73 kJ of heat added:

$$CH_4 = C + 2H_2$$

It should be mentioned that this reaction produces carbon black, which is difficult to recycle.

- extracting CO and CO_2 from the exhaust gas of burnt methane;
- recovering hydrogen from methane by the two-stage decomposition process given by

1. $CH_4 + H_2O = CO + 3\,H_2$
2. $CO + H_2O = CO_2 + H_2$

The carbon monoxide (CO) reacts exothermally (42 kJ of heat released) with water, generating hydrogen and CO_2, which have to be removed.

CO_2 and CO produced by the last two processes could be re-used in the manufacture of methane from hydrogen, thus completing a cycle – the carbon here acts as a hydrogen 'carrier'.

Methane is already widely distributed in the form of natural gas and can be used directly as a fuel with relatively little CO_2 emission during combustion, especially if enriched with hydrogen; its use does not require any new technology.

Ammonia is probably more attractive than methane as an example of a system utilising a nitrogen 'carrier', which would not incur any return transformation costs, as it would be re-cycled through the atmosphere. Gaseous ammonia is prepared from hydrogen by direct combination with nitrogen, under 200–500 bar pressure and at 720 K in the presence of a catalyst, using the Haber process:

$$N_2 + 3H_2 = 2NH_3 \quad \text{exothermic } \delta H = -90\ \text{kJ}$$

High temperature and pressure are essential to obtain a sufficiently high reaction rate. The reaction results in an overall reduction in volume of the reactants. Fuel is collected by the liquefaction of gaseous ammonia at its boiling point of 240 K. The reaction is exothermic and generates 90 kJ heat. Heat would need to be supplied to the reverse reaction – decomposition of ammonia to form hydrogen and nitrogen – which is endothermic and would require an input of the same amount of heat (90 kJ). At 400 °C and 10 bar, ammonia is dissociated to the extent of 98%.

Dissociating ammonia into hydrogen and nitrogen is a simple process of passing it, at relatively low pressure, through a hot tube. However, it is necessary to establish to what extent the nitrogen would need to be separated from the hydrogen before consumption.

The potential advantages of storing ammonia, relative to storing hydrogen, are

- safer storage;
- energy density is higher; and
- easier to liquefy.

Such a system will be attractive where the site of energy generation is remote from its place of consumption, since it is cheaper to transport ammonia than hydrogen.

Methylcyclohexane, formed from the direct hydrogeneration of toluene, is a liquid that could be used in the same way as petroleum products. For example, it could be consumed in engines instead of petrol or diesel fuels, but this would be unattractive since the reduction in the emission of CO and CO_2 would be hardly noticeable relative to petrol. It may also be used to regenerate hydrogen, but in this case the toluene 'carrier' would need to be returned for rehydrogeneration, thus reducing significantly the mass transport efficiency.

8.4 Storage containment for hydrogen

The options exist for the storage of hydrogen:

- compressed gas;
- chemical compounds;
- liquid hydrogen; and
- metallic hydrides.

The relative capital and other specific costs for these alternatives are given in Table 8.3.

Table 8.3 Relative costs of different hydrogen containments

Containment concept	FeTi hydride	Liquid H_2	Gaseous H_2 50 bar
Relative capital cost	15.09	54.09	13.72
Relative annual operating cost	0.3	23.2	3.42
Storage cycle relative cost per volume (N m^3)	2.54	20.27	3.0
Storage cycle relative cost per kWh	0.82	6.72	1.0

Since the cost of liquefaction is considerable, the most attractive concept for the bulk storage of hydrogen produced from substantial non-oil-based primary energy sources is compressed gaseous hydrogen in underground caverns, where it can be stored in a similar way to natural gas. The relevant technology has been discussed in Chapter 7. The high diffusivity of gaseous hydrogen will probably have little effect on leakage since most rock structures are sealed in their capillary pores by water.

Alternative systems for using hydrogen include its chemical combination with other elements to form compounds that are more amenable to storage. Table 8.1 lists the properties of a range of possible compounds, such as methane and ammonia, containing chemically bonded hydrogen. They are compared with hydrogen itself and were selected as representative of the three different approaches discussed in Section 8.3.

Liquid hydrogen has a mass energy density three times greater than oil; its usage is particularly attractive for heavy surface transport and aircraft, allowing improvement in payloads and extending range. Because of its low density, liquid hydrogen may be less attractive than other materials for its energy storage density on a volumetric basis. Ammonia and methane, as liquids, would appear to be the most efficient materials (see Table 8.1) for storing hydrogen on both a weight and a volume basis.

To store liquid hydrogen and other cryogenic fluids, as well as gas under pressure, reliable low-mass storage vessels will be necessary.

Interest in the use of low-mass composite materials to contain cryogenic fluids arose around 1960 with the US Apollo programme. It was decided then to line the composite vessel with a metallic permeability barrier, to prevent the loss of hydrogen and also to contain the cryogenic liquid in the event of mechanical damage, which could arise from the low-temperature fatigue characteristics of composite materials. While the room temperature fatigue properties of composite materials are excellent, there are uncertainties about their cryogenic performance, particularly in a low-temperature reactive environment such as hydrogen and oxygen. At the Rutherford Appleton Laboratory, Dr D. Evans has shown that the permeability to hydrogen of some modern advanced composite materials is low, and therefore it is technically possible to use storage vessels made of these materials without including barrier layers

Other methods for 'fixing' hydrogen in order to store it more effectively would include the preparation of hydrides or physical absorption on materials such as zeolites ('molecular sieves'). Liquid hydrogen occupies approximately four times the volume of hydrocarbon fuels, so the concept of hydrogen dissolved in metals, the so-called hydride concept, has been proposed. The idea is that a number of metal hydrides, such as $LaNi_5$, TiFe and Mg_2Ni, can absorb and release hydrogen at low pressures and temperatures with small losses.

8.5 The hydride concept

Hydrogen molecules may be dissociated to atoms when they interact with a metal or alloy. These atoms may be absorbed in the metal unless the limit of solution is reached and equilibrium hydrogen pressure is established. When all the material has changed into hydride, the pressure may be increased steeply again. Depending on the number of existing phases, pressure levels will rise and the process will be repeated. The excess pressure at which hydrogen is released from the hydride is orders of magnitude less than that of gaseous hydrogen stored in the pressure vessel because of the dissociation of molecules into atoms and the bonding in the metallic phases. For the same reason, the volume density of hydrogen in hydrides is greater than in the gaseous and liquid forms. An illustration of concentration pressure isotherms at a given temperature is shown in Figure 8.2.

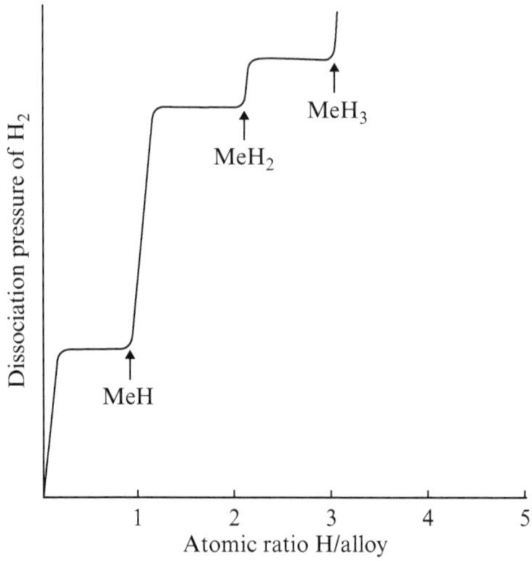

Figure 8.2 Relative pressure isotherms for different H_2 concentrations in a hydride (T = constant)

The aim is to select a hydride that can be thermally decomposed in a reversible manner so that hydrogen may be withdrawn or replenished from/to the vessel when necessary. A suitable hydride store has to have the following features:

- high hydrogen content per unit mass of metal;
- low dissociation pressure at moderate temperatures;
- constancy of dissociation pressure during the decomposition time;
- safe on exposure to air; and
- low cost.

The exothermic chemical reaction of hydride formation from metals (Me) and hydrogen (H_2) may be given as follows:

$$H_2 + \text{Me} \underset{\substack{\text{Discharging} \\ \text{(heat added)}}}{\overset{\substack{\text{Charging} \\ \text{(heat released)}}}{\Leftrightarrow}} \text{hydride} + \text{heat}$$

Low-temperature FeTi hydride, with a low energy requirement for hydrogen release, has been developed by Brookhaven National Laboratory, Long Island, NY, USA. Magnesium-based high-temperature hydride has attracted interest because it is a readily available and rather cheap metal. Both materials release hydrogen endothermically, thus creating no safety problems. The mass energy densities (W_m) of hydrides based on Ti and Mg are

$FeTiH_{1.7} \rightarrow FeTiH_{0.1}$	$1.856 \times 10^6 J/kg$
$Mg_2NiH_4 \rightarrow Mg_2NiH_{0.3}$	$4.036 \times 10^6 J/kg$
$MgH_2 \rightarrow MgH_{0.005}$	$9.198 \times 10^6 J/kg$

A combination of hydrides with different plateau pressures offers a variety of applications involving both hydrogen storage and thermal storage.

When hydrides are used as hydrogen-storing vessels for heat engines or for domestic heaters, the waste heat from them can be returned back to the hydride. If the amount of waste heat is less than that needed for hydrogen release, then the waste heat can be stored as thermal energy in the metal hydride (which acts as a thermal storage device). A combination of hydrides with different release temperatures may be used for different applications such as heat pumps, central heating, air conditioning, and even for renewable-storage systems. This possibility has led to the so-called 'hydride concept' proposed by Daimler-Benz in Germany.

This concept is based on the separation in time and location between the combustion process and the release of the waste heat produced during this process. It is possible because:

- The heat transfer per minute can be controlled by varying the rate at which hydrogen is withdrawn from or added to the hydride.
- The pressure/temperature characteristics of the hydride determine the temperature.

Systems that comprise high- and low-temperature hydrides work as follows: the exhaust gas from the hydrogen-fired machine passes over high-temperature hydride, releasing hydrogen, which is then pumped to the low-temperature hydride. The exhaust then passes over the low-temperature hydride, thereby releasing hydrogen to the hydrogen-fired machine and increasing the temperature of the hydride. Cooling water from the machine has to be used for hydrogen dissociation.

Figure 8.3 gives a survey of the hydride energy concept, in which hydrides can be used for mobile, stationary or electrochemical storage, or hydrogen/deuterium separation.

An alternative to gaseous hydrogen as an energy carrier is to bond it with nitrogen from the air to produce ammonia, which can be easily liquefied, transported and used (directly or indirectly) as a clean and renewable fuel. The toxicity of ammonia is one of the main issues holding back an ammonia economy.

The methanol economy is a synthetic fuel production energy plan that may begin with hydrogen production. Hydrogen in a full 'hydrogen economy' was initially suggested as a way to make renewable energy in non-polluting form,

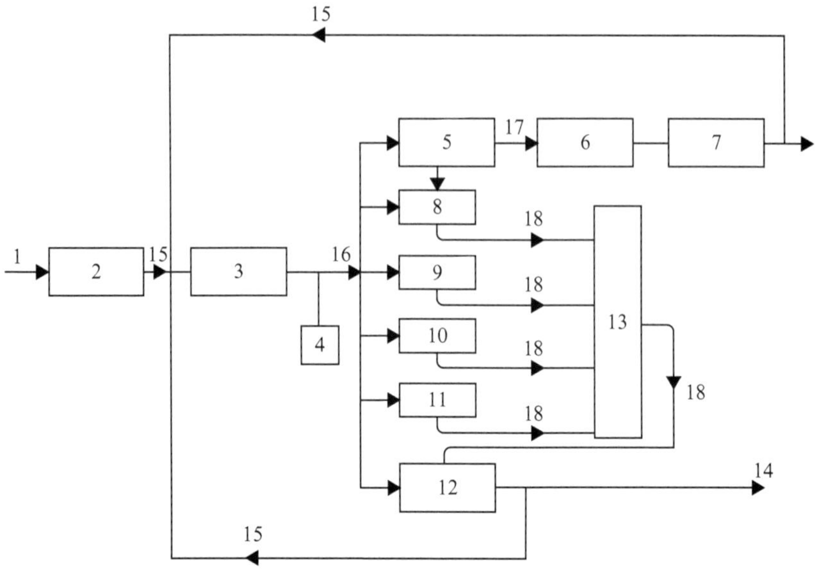

1. Primary energy sources – nuclear, hydro or renewable
2. Base power plant
3. Electrolyser
4. Hydrogen storage
5. H_2/D_2 separation from TiNi hydrides
6. Natural uranium reactor
7. TiNi battery
8. Hydride storage for automobile vehicles
9. Synthetic chemical and fuels production
10. Heat chemical processes
11. Domestic heaters and hydride domestic air conditioner
12. Local thermal power stations
13. Waste heat recovery
14. Customers
15. Electric current
16. H_2
17. D_2
18. Heat

Figure 8.3 The hydrogen storage concept

available to automobiles that are not all-electric. However, a theoretical alternative to direct elemental hydrogen use in vehicles would address the same problem by using centrally produced hydrogen immediately, to make liquid fuels from a CO_2 source. Thus, hydrogen would be used captively to make fuel and would not require expensive hydrogen transportation or storage. To be greenhouse-neutral, the source for CO_2 in such a plan would need to be from air, biomass or CO_2 that would otherwise be scheduled to be released into the air from non-carbon-capture fuel-burning power plants (of which there are likely to be many in the future, since economic carbon capture and storage is site-dependent and difficult to retrofit).

Many of the hybrid strategies using captive hydrogen to produce other more easily usable fuels might be more effective than hydrogen production alone. Short-term energy storage (meaning the energy is used not long after it has been captured) may be best accomplished with battery or even ultracapacitor storage. Longer term energy storage (meaning the energy is used weeks or months after capture) may be better done with synthetic methane or alcohols, which can be stored indefinitely at relatively low cost, and even used directly in some type of fuel cells, for electric vehicles. These strategies dovetail well with the recent interest in Plug-in Hybrid Electric Vehicles, or PHEVs, which use a hybrid strategy of electrical and fuel storage for their energy needs.

Chapter 9
Electrochemical energy storage

9.1 General considerations

The most traditional of all energy storage devices for power systems is electrochemical energy storage (EES), which can be classified into three categories: primary batteries, secondary batteries and fuel cells. The common feature of these devices is primarily that stored chemical energy is converted to electrical energy. The main attraction of the process is that its efficiency is not Carnot-limited, unlike thermal processes. Primary and secondary batteries utilise the chemical components built into them, whereas fuel cells have chemically bound energy supplied from the outside in the form of synthetic fuel (hydrogen, methanol or hydrazine). Unlike secondary batteries, primary batteries cannot be recharged when the built-in active chemicals have been used, and therefore strictly they cannot be considered as genuine energy storage. The term ‘batteries’, therefore, will only be applied for secondary batteries in this chapter.

Batteries and fuel cells comprise two electrode systems and an electrolyte, placed together in a special container and connected to an external source or load. These two electrodes, fitted on both sides of an electrolyte and exchanging ions with the electrolyte and electrons with the external circuit, are called the anode ( ) and cathode (+), respectively (see Figure 9.1).

The anode is defined as the oxidising electrode, that is, the electrode that is sending positive ions into the electrolyte during discharge. When supplying positive charges to the electrolyte, the anode itself becomes negatively charged and therefore may be considered as an electron source for the external circuit. At the same time, the cathode consumes electrons from the external circuit and positive ions from the internal circuit. To maintain electric current in the external circuit, electrons have to be produced at the anode and used up at the cathode. Since no chemical process can generate electric charge, the transport of charge in the electrolyte (in the form of ions between the electrodes) has to take place at the same time.

It should be mentioned that no electron conductivity must take place as, in that case, the battery will discharge itself through a short-circuit.

The electromotive force (EMF), of a battery, which initiates the electric current, is the difference between the electric potential of the electrodes. The terminal voltage V equals the EMF minus the voltage drop in the battery due to its internal

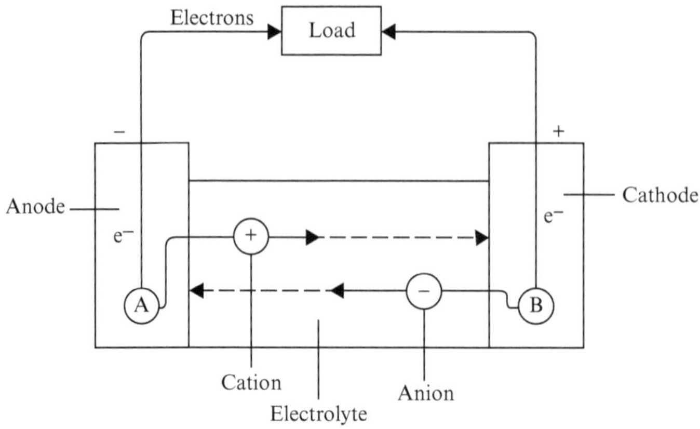

Figure 9.1 Electrical energy source during discharge

resistance *R*, which contains frequency and time-dependent components associated with the electrolytic processes, ohmic resistance against the charge transport in the entire internal circuit components, an external load dependence component and the remaining energy contents of battery component. In other words, the internal resistance can be described in the form of a rather complicated impedance. The smaller the value of internal resistance the lighter is a battery's turnaround efficiency, since there is a linear dependence between thermal losses in a battery and its internal resistance.

High reaction rate and good transport conditions will lead to a substantial decrease in the irreversible thermal losses in any electrochemical battery. Both factors can be met by working at high temperatures and with chemically active electrodes. In both cases, the electrolyte will be a limiting factor owing to problems of stability and transport properties.

A common solution is to use different types of aqueous electrolytes, but unfortunately they can only work at high temperatures under substantial pressure, therefore requiring special vessels. The use of ceramic materials, which have a suitably high specific conductivity for ions able to take part in the electrochemical process, is a promising possibility. The development of these so-called solid-state ion conductors has contributed to a breakthrough in battery technology.

9.2 Secondary batteries

Rechargeable electrochemical batteries have a long history of application in electrical power systems. At the beginning of the century, diesel engines used for generation in small, local DC power systems were usually shut down at night and the demand was met by lead-acid batteries that had been charged during the day. These batteries were also used in several US towns to feed DC electricity to electric street cars during rush hour. The growth of large centralised AC power systems and cheap coal- and oil-generated electricity relegated batteries to emergency standby duty for DC auxiliaries. In this application, they are still invaluable today; over 100 MW of battery power capacity is currently installed for standby duty throughout the National Power and PowerGen power companies in the United Kingdom.

Let us consider the cell reactions of the most popular battery systems, together with a brief description of their main problems.

The battery market is still dominated by the lead-acid battery invented by Plante in 1859 (see Figure 9.2). It is the oldest chemical storage device. The main drawbacks of these batteries are comparatively low energy density, long charge time and the need for careful maintenance. Nearly half the weight of a lead-acid battery is occupied by inert materials, for example grid metal, water, separators, connectors, terminals and cell containers, and attempts to reduce the weight, to increase the energy density, have involved the use of low-density materials. The use of carbon fibre in the positive electrode system has led to a reduction in weight and also increased the power capability of lead-acid cells. The increasingly widespread use of alloy negative electrodes, based on Pb/Ca, reduces H_2 evolution and water loss, thus minimising the requirement for maintenance.

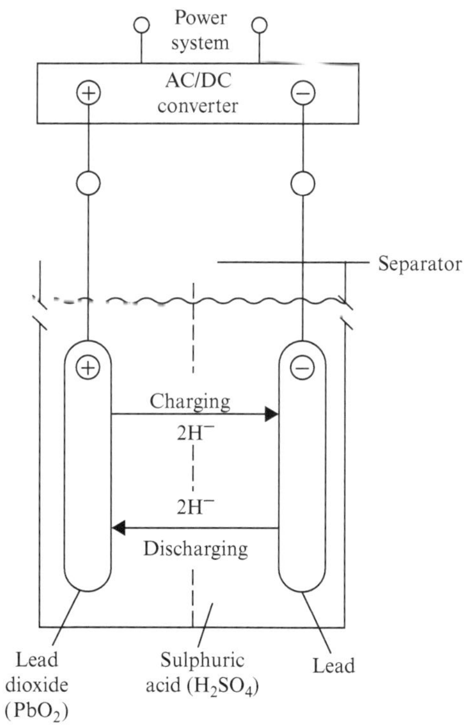

Figure 9.2 Lead-acid storage battery

The battery consists of alternate pairs of plates, one lead and the other lead coated with lead dioxide, immersed in a dilute solution of sulphuric acid that serves as an electrolyte. Here only one pair of plates, or electrodes, is shown. During discharge, both electrodes are converted into lead sulphate ($PbSO_4$). Charging restores the positive electrode to lead dioxide (PbO_2) and the negative electrode to metallic lead. The performance of such a battery deteriorates gradually because of irreversible physical changes in the electrodes; ultimately, failure occurs between several hundred and 2000 cycles, depending on battery design and duty cycle.
Reaction at positive electrode

$$Pb_2SO_4 + 2H_2O \underset{\text{discharge}}{\overset{\text{charge}}{\rightleftarrows}} PbO_2 + H_2SO_4 + 2H_2 + 2e^-$$

and reaction at negative electrode

$$PbSO_4 + 2H_2 + 2e \underset{\text{discharge}}{\overset{\text{charge}}{\rightleftarrows}} H_2SO_4 + Pb$$

are given as follows:

$$\overset{\text{Anode}}{Pb} + 2H_2SO_4 + \overset{\text{Cathode}}{PbO_2} \underset{\text{Charge}}{\overset{\text{Discharge}}{\rightleftarrows}} \underset{\text{Anode}}{PbSO_4} + 2H_2O + \underset{\text{Cathode}}{PbSO_4}$$

As mentioned previously, the lead-acid battery has been the popular choice as a power source both for traction and stationary purposes. Large 1–5 MW lead-acid installations also have an established history in submarine use, routinely achieving 2000 cycles.

Industrial development on alkaline electrolyte batteries, such as Ni-Zn and Fe-Ni, aims to produce improved energy storage systems for traction applications in the foreseeable future. The Ni-Zn battery is currently considered as a possible medium-term EV system. It is analogous to the much more expensive Ni-Cd battery. The cell reaction may be written as follows:

$$\overset{\text{Cathode}}{2NiOOH} + 2H_2O + \overset{\text{Anode}}{Zn} \underset{\text{Charge}}{\overset{\text{Discharge}}{\rightleftarrows}} 2Ni(OH)_2 + Zn(OH)_2$$

The Ni-Zn cell's main drawbacks are its short life cycle, separator stability, temperature control, high cost and mass production problems. Its short life cycle is the result of the high solubility of the reaction products at the zinc electrodes. Re-deposition of zinc during charging causes the growth of dendrites, which penetrate the separators of the battery and cause an internal short-circuit, and also the redistribution of active material. Possible solutions to the problems of dendritic growth and slumping of the zinc electrode in alkaline electrolyte are the use of electrode and electrolyte additives and penetration-resistant separators. Attempts have been made to suppress the growth of zinc dendrites during the charge mode by vibrating the zinc electrode.

The robust, long-life nickel-iron battery has become popular again after a period of 70 years during which its design remained practically unchanged from that of Thomas Edison; the Westinghouse Corporation has developed an EV system employing fibrous electrodes and electrolyte circulation.

The Ni-Fe battery is an alkaline storage battery using KOH as the electrolyte. The cell reaction is

$$\overset{\text{Cathode}}{2NiOOH} + 2H_2O + \overset{\text{Anode}}{Fe} \underset{\text{Charge}}{\overset{\text{Discharge}}{\rightleftarrows}} 2Ni(OH)_2 + Fe(OH)_2$$

The main drawback of Ni-Fe batteries for electric vehicle applications has been their rather low energy density, although recent Japanese and Swedish developments are fast making it as attractive as the Ni-Zn system in this respect.

The battery, however, also has a low peaking capability, and other drawbacks are low cell voltage, which leads to more cells being needed for a given battery voltage requirement, and the low hydrogen overvoltage of the iron electrode, which results in self-discharge and low cell efficiency.

Nickel battery systems will all benefit from improvement of electrochemical impregnation of the active material. The long life cycle claimed for these metal–metal oxide cells will lead to good economic characteristics even though they possess high initial capital costs.

The iron-air battery comprises an anode made of iron and a cathode taking oxygen from the air. The cell reaction between oxygen from the air and iron may be given as follows:

$$\overset{\text{Anode}}{Fe} + H_2O + \overset{\text{Cathode}}{1/2O_2} \underset{\text{Charge}}{\overset{\text{Discharge}}{\rightleftarrows}} Fe(OH)_2$$

The battery has a high self-discharge level for the iron electrode at low temperatures, low charge efficiency and its power capacity is limited to 30–40 W/kg.

The zinc-air battery has highly concentrated KOH electrolyte and the electrochemical reaction is between oxygen from the air and zinc metal. The cell reaction can be written as

$$\overset{\text{Anode}}{Zn} + \overset{\text{Cathode}}{1/2O_2} \underset{\text{Charge}}{\overset{\text{Discharge}}{\rightleftarrows}} ZnO$$

As with the iron-air system, the zinc-air couple has a poor overall charge–discharge efficiency owing to polarisation losses associated with the air electrode. Re-deposition of zinc during charging leads to problems of electrode shape change and dendritic growth. All these lead to zinc-electrode instability, which is the main technical problem in the development of this type of battery. The zinc oxide formed during discharge dissolves in the electrolyte to give zincate ions. Cell lifetime and efficiency are the main economic problems, which were greatly changed after the introduction of high-speed circulation of electrodes, which aimed to produce zinc deposits.

Iron-air and zinc-air batteries suffer crucially from the low efficiency of the air electrode, which is also susceptible to damage during charge at useful current densities. Nevertheless the iron-air battery is being developed for traction purposes by Westinghouse and the Swedish National Development Corporation who use wet-pasted carbon and dual-layer nickel materials, respectively, for the air electrode, both employing silver as a catalyst.

The prospect for Zn-Cl batteries appears commercially promising since $ZnCl_2$ cells will overcome the replating problem of zinc by using an acid electrolyte containing additional chloride salts. The cell reaction is given by

$$\underset{\text{Anode}}{Zn} + \underset{\text{Cathode}}{Cl_2} .6H_2O \underset{\text{Charge}}{\overset{\text{Discharge}}{\rightleftarrows}} ZnCl_2 + 6H_2O$$

Energy Development Associated has proposed a module with an energy capacity of 1.8×10^8 J, which stores the chlorine as chlorine hydrate, $CI_2.6\,H_2O$ below 9 °C. Even after refrigeration and pumping losses, the overall efficiency is projected to be approximately 0.65, which is comparatively good. The gradual oxidation of the porous graphite $C1_2$ electrode leads to CO_2 generation in the battery. Migration of Cl_2 to the Zn electrode during charge, which also tends to evolve H_2, leads to a decrease in efficiency and is a serious problem.

The main problem with the Zn-Cl hydrate battery is that the system is too complicated. The auxiliary equipment for refrigerating, heating and storing the chlorine creates many difficulties and places severe weight and volume penalties on the system. For example, a 10^{12} J storage device will require 60 tonnes of CI_2. The safety problems of Cl_2 storage are perhaps the greatest drawback for this type of cell.

The Zn-Br_2 cell, using aqueous $ZnBr_2$ electrolyte, can avoid storage of free halogen by using complexing agents to form Br_3^- and Br_5^-. General Electric has developed a laboratory cell with carbon electrodes and a solid polymer membrane, which prevents the bromine-carrying ions reaching the zinc electrode. A system using a microporous separator and long-life RuO-catalysed titanium electrodes is under development by Could, USA. These systems will probably withstand over 2 000 charge–discharge cycles.

Advanced batteries, in particular the sodium-sulphur (Na-S) couple using a solid-state electrolyte and the Li-S couple using a fused salt electrolyte, are still under development.

The Na-S battery is one of the most advanced high-temperature battery concepts. A battery of 10^6 J energy capacity for EV applications is under way at Chloride Silent Power, UK, in collaboration with GE, USA. This system is also under development by British Rail in the UK, Brown Boveri Co. in Germany and

Ford in the USA. All these concepts use a tube of Na-conducting beta-alumina ceramic as an electrode, which is able to conduct sodium ions with molten Na on one side, and a sodium polysulphide melt on the other contained in a carbon felt for current collection (see Figure 9.3).

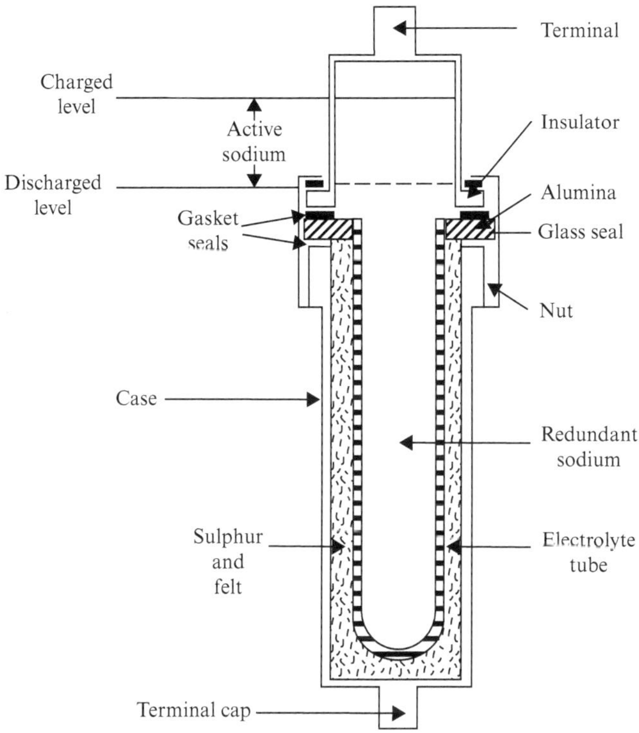

Figure 9.3 *Schematic diagram of the Na-S battery that uses a sodium beta-alumina solid electrolyte as the separator between liquid electrodes (sodium anode and sulphur cathode). The operation temperature is 300–400 °C*

An isomorph of aluminium oxide with the so-called beta-alumina crystal structure has very high sodium ion mobility above room temperature, as discovered by Weber and Kummer. Beta-alumina has a layer structure; four layers of closely packed oxygen atoms contain fourfold and sixfold coordinated aluminium in the same atomic arrangement as spinel, $MgAl_2O_4$. However, the formula of beta-alumina, $Na_{1+x}Al_{11}O_{17+1/2x}$, is not compatible with an infinitely extending spinel structure, as there is a relative deficiency in aluminium and every fifth layer of oxygen atoms is only one-quarter filled. The sodium ions are found in this relatively empty layer. When the electrical conductivity of beta-alumina was measured at 300 °C, it became clear that the charge carriers were exclusively sodium ions and not electrons; the application of such materials to secondary batteries was quickly realised. In the Na-S battery patented by Ford, rather than solid electrodes separated by a liquid electrolyte (as in the conventional lead-acid battery), sodium beta-alumina is used as a solid electrolyte, conducting sodium ions between liquid electrodes of sodium metal and sulphur. The cell voltage (2.08 V) is derived from the chemical reaction between sodium and sulphur to produce sodium polysulphide. The theoretical energy density of these batteries is approximately 2.7×10^6 J/kg, much higher than the 0.61×10^6 J/kg of the lead-acid battery.

The reaction can be given by

$$\underset{\text{Anode}}{x\text{Na}} + \underset{\text{Cathode}}{y\text{S}} \underset{\text{Charge}}{\overset{\text{Discharge}}{\rightleftharpoons}} \text{Na}_x\text{S}_y$$

The first problem in developing beta-alumina electrolytes for Na-S batteries includes research into composition, phase equilibria and influence of micro-structure on the mechanical and chemical properties of beta-alumina. The second problem is cracking of the ceramic tube and insulating seals, together with corrosion at the sulphur electrode, leading to shortened cycle life. A third problem is related to the overall design of the cells and the formation of insulating sulphur layers at high charging current densities. Despite some progress, much remains to be done, especially in the field of ceramic production technology, but if these problems can be overcome, the Na-S battery could be very attractive for power system use.

The Li-S battery comprises liquid lithium, sulphur electrodes and electrolyte of molten LiCl-KCl eutectic at an operating temperature in the 380–450 °C range. The cell reaction is given as follows:

$$\underset{\text{Anode}}{2\,\text{Li}} + \underset{\text{Cathode}}{\text{S}} \underset{\text{Charge}}{\overset{\text{Discharge}}{\rightleftharpoons}} \text{Li}_2\text{S}$$

Highly corrosive liquid lithium attacks the ceramic insulators and separators and shortens the cell's life. Efficiency is not very high because of self-discharge caused by lithium dissolving in the molten LiCl-KCl electrolyte.

The use of lithium-aluminium alloys and iron sulphide as electrodes has led to the development of more efficient Li-S cells with reasonable energy densities. The reactions in the cell can be written as follows:

$$\underset{\text{Anode}}{4\text{LiAl}} + \underset{\text{Cathode}}{\text{FeS}_2} \underset{\text{Charge}}{\overset{\text{Discharge}}{\rightleftharpoons}} \text{Fe} + 2\text{Li}_2\text{S} + 4\text{Al}$$

or

$$\underset{\text{Anode}}{4\text{LiAl}} + \underset{\text{Cathode}}{\text{FeS}} \underset{\text{Charge}}{\overset{\text{Discharge}}{\rightleftharpoons}} \text{Fe} + \text{Li}_2\text{S} + 2\text{Al}$$

The Li-FeS battery studied at Argonne National Laboratory, USA, in collaboration with Eagle-Picher and Gould, comprises multiples of vertical Li-Al (negative) and FeS (positive) electrodes separated by boron nitrite felt or MgO_2 powder. It employs a KCl-LiCl high-temperature eutectic melt as electrolyte, for which a vacuum multifoil container vessel is needed.

A promising system is solid Li_4Si coupled with $TiS_2/Sb_2S_3/Bi$. Fine Al_2O_3 powder, with dispersed LiI, is used as the solid electrolyte. A voltaic pile would be the best construction for this system with thin-film technology providing the possibility of low-cost production.

The Li-TiS_2 cell has a lithium metal anode and an intercalation cathode of TiS_2. The insertion of lithium ions between adjacent sulphur layers leads to the electrochemical reaction:

$$\underset{\text{Anode}}{x\mathrm{Li}} + \underset{\text{Cathode}}{\mathrm{TiS_2}} \underset{\text{Charge}}{\overset{\text{Discharge}}{\rightleftarrows}} \mathrm{Li}_x\mathrm{TiS_2}$$

The intercalation reaction proceeds without any change in the host matrix except for a slight expansion of the *c*-axis. The charge–discharge process is perfectly reversible and operates at room temperature. Although the battery is still not commercially available, it attracts considerable attention because of its high energy density, long life, a leak-proof seal and the ability to indicate the level of its charge.

Redox cells are of considerable promise for power system use, including the rechargeable flow cell $TiCl_3TiCl_4FeCl_3FeCl_2$. This cell comprises two electrolyte compartments divided by a membrane. During the discharge mode, $FeCl_3$ is reduced to $FeCl_2$, while $TiCl_3$ is oxidised to $TiCl_4$. Chlorine ions (Cl^-) can penetrate the membrane between the electrolyte compartments, which preserves the electroneutrality of the cell. The redox flow cell operates at ambient temperature and the overall efficiency is 70%. Unlike conventional batteries, there are no apparent cycle life limitations due to changes in the active electrode materials.

A solid-state battery concept is very attractive for power system applications. Containment of reactive materials would be an integral part of the battery itself and corrosion should be eliminated, thus increasing the total life.

Since the middle of the first decade of the 21-st century newer lithium-ion batteries technologies have been rapidly developed. These provide significant utility scale load leveling capabilities. The usage of lithium-ion batteries in electric cars improves significantly their performance and is leading to the large scale introduction of electric and hybrid cars to consumers over the next decade.

9.3 Fuel cells

Fuel cells, distinguished from other secondary batteries by their external fuel store, have an even longer history than the lead-acid battery. The first H_2-O_2 fuel cell was demonstrated in principle by the English lawyer W.R. Grove in 1839, although the bulk of the fuel cell's development has been in the last 40 years and the major application has been in the space industry.

Most research activity in secondary batteries worldwide is concerned with materials research for advanced batteries, in particular materials for solid solution electrodes and for solid electrolytes. It should be noted that solid electrolytes are equally applicable to electrochemical fuel cells.

The mobile positive ions in the electrolyte may be metal ions, produced by the metal anode supplying electrons to the external circuit in the battery, while in a fuel cell they may be H^+ (H_3O^+), produced by the hydrogen supplied to the anode. A very simplified transport model is shown in Figure 9.4.

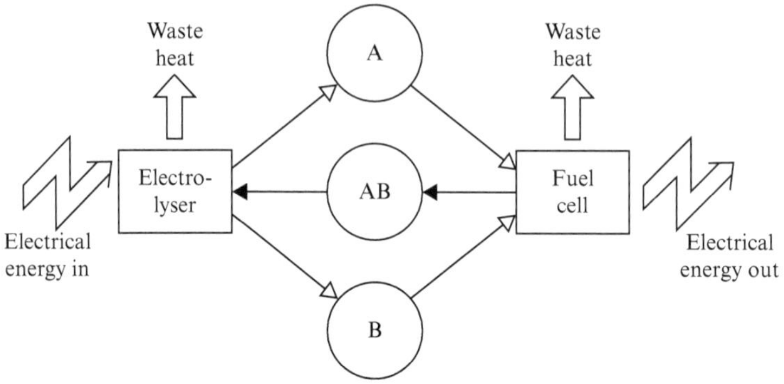

Figure 9.4 Schematic diagram of an electrochemical fuel cell

Let us consider an electrochemical concentration cell with a reaction given by

$$\underset{\text{electrode 1}}{X(a_1)} \rightleftharpoons \underset{\text{solid electrolyte: conducting X as an ion}}{} \rightleftharpoons \underset{\text{electrode 2}}{X(a_2)}$$

The EMF of such a cell may be calculated with the help of the Nernst equation:

$$E = RT\ln(a_1/a_2)/(nF)$$

where n is the number of electrons needed to get one atom or molecule of X into its ionic form in the electrolyte, and a_1 and a_2 are the activities at electrodes 1 and 2.

We can make use of such a cell in the following ways:

(i) If $a_1 > a_2$ and X is continually added on the left, and removed on the right, we have a source of energy – a concentration fuel cell.
(ii) If we apply a greater voltage than E in the opposite sense, we can drive X from one side to the other – hence we have an ion pump or an electrolyser.

As a fuel cell is an electrochemical cell that can continuously change the chemical energy of a fuel and oxidant to electrical energy with high efficiency, it is not surprising that a variety of synthetic fuels have been proposed, such as hydrogen, methanol, ammonia and methane.

H_2-O_2 fuel cells have attracted special interest mainly because of the concept (previously mentioned) of hydrogen economy.

The essential functions of a fuel cell are

1. the charging (or electrolyser) function in which the chemical AB is electrochemically decomposed to A and B;
2. the storage function in which A and B are held apart; and

3. the discharge (or fuel-cell) function in which A and B are reunited, with the simultaneous generation of electricity.

In secondary batteries, the electrolyser and fuel-cell functions are combined within the same cell. This is a convenient arrangement but is not essential.

These fuel cells have the overall reaction

$$H_2 + \frac{1}{2}O_2 \rightarrow H_2O$$

and are attractive because of their high energy density, lack of pollution and high cell efficiency. As seen from the overall equation, a good hydrogen or oxygen conductor has to be used as the electrolyte.

The $H_2{-}O_2$ fuel cell system usually comprises a separate electrolyser and a fuel cell itself with the gases stored at 25 atm in underground tanks. There is a clear advantage in combining the fuel cell and electrolyser in the same installation. Development in this direction is perhaps promising in relation to the solid polymer electrolyte NAFION, proposed by GE, USA. The related $H_2{-}Cl_2$ and $H_2{-}Br_2$ systems appear more promising still and a laboratory $H_2{-}Br_2$ cell (using a NAFION membrane) has demonstrated good efficiency at a useful current density.

A particular feature of H_2-halogen fuel cells, and also redox systems, is that the capital cost of extra energy storage capacity is independent of the cost of electrodes but is very clearly related to the electrolyte and battery vessel costs. These can be quite low, particularly for the $H_2{-}Cl_2$ system, and, if there is substantial reliance on renewable energy sources (i.e. more than 20% for the United Kingdom), such battery storage could be attractive for application in the power system. However, their relatively large size would probably create siting problems and mean that few could be placed close to load centres, thus the desired transmission savings would not be available.

9.4 Storage unit assembly

As can be seen from Table 9.1, cell voltages V_c are generally quite low. Cell current is constrained by voltage and inner resistance R_c. To create a powerful storage system, special series–parallel cell connection is needed. This strategy was first developed for large load-levelling installations for power systems but also offers similar benefits for present-day applications.

If storage rated power capacity P_s and voltage V_s are known, the number of parallel connected cells n_p in a module and the number of series-connected modules n_s may be calculated from

$$n_s = \frac{V_s}{V_c}$$

$$n_p = \frac{I_s}{I_c} = \frac{P_s R_c}{V_s V_c}$$

It is necessary to add a certain number of redundant cells to each module to minimise the impact of cell failure on store maintenance; one redundant cell in five

Table 9.1 Electrochemical energy storage

Couple	**Pb-PbO$_2$ (lead-acid)**	**NiO-Cd (nickel-cadmium)**	**NiO-Fe (nickel-iron)**	**Na-S (sodium-sulphur)**	**Li-S (lithium-sulphur)**	**Li-TeCl (lithium-chlorine)**	**Zinc-air**	**Iron-air**	**H$_2$–O$_2$ (hydrogen-oxygen)**
Electrolyte	H_2SO_4	KOH	KOH	solid beta-Al_2O_3	Molten halide eutectic	Molten alkali halide	KOH	KOH	
State of development	1–2 MW installation	Wide range of small batteries	Proven, traction	Prototype batteries (10 kWh)	LiCl/LiI/KI commercial production	eutectic commercial production (2–3 kWh)	Commercial production	Laboratory cells	Commercial electrolysers
Operation temperature (°C)	−20, +50	−30, +50	10, 50	300, 400	430 500	460	0, +60	0, +50	127 227
Life cycles	500	2 000	2 000	2 000	200	1 000	100	200	2 000
Overall efficiency	75%	70%	60%	75%	75%	68%	55%	40%	42%
Recharge time (h)	5–8	4–7	4–7	7–8	5	7	5–8	4–5	
Charge/discharge period (h)	5/7	7/5	7/5	7/5	7/5	7/5	–	–	7/10
Specific energy (kJ/kg h)	86	100	144	432	500	–	290	180	120 000
5 h	144	110	200	504	–	–	360	290	120 000
Energy density (MJ/m^3 h)	252	216	360	612	–	–	290	290	8 960
Specific power, peak	120	300	440	240	200	–	100	60	–
W/kg sustained	250	140	220	120	140	–	–	50	–
Voltage V open circuit	2.05	1.35	1.37	2.1–1.8	1.9	3.1	1.65	1.27	1.2
2 h discharge	1.9	1.2	1.2	1.7–1.4	1.3	2.9	1.2	0.7	–
Relative capital cost, p.u.	1.0	2.0	1.25	0.65	0.3–0.5	0.43	3	0.92	1.15

is recommended, i.e. 20% spare capacity. This level of redundancy reduces the expected number of module repair operations to one per installation per ten years, compared with 60 with no spare cells and 8 with 10% spare capacity.

The cells in each module have to be arranged in two rows and are supported by a conductor beam located between the two rows. Each cell has to be attached to the beam with the help of a cell-shaped band welded to the beam. A number of bands have to be placed at the end of each module to receive 'booster' cells. These are the means of repairing a module in situ should this be necessary at any time over the battery's lifetime. This maintenance concept minimises battery repair time since defective modules do not need to be removed for change, and considerably simplifies the central store support frame, which comprises columns, longerons and lateral interties. The module support beams are cradled on the longerons and are electrically isolated from the main frame. The central store comprises interconnected cell modules, intermodule busbars, module separators, the support frame and an insulated enclosure. This is shown conceptually in Figure 9.5.

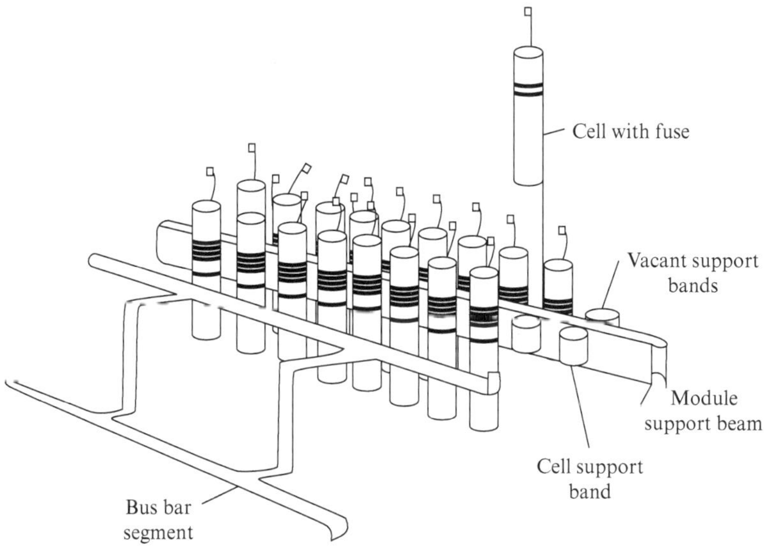

Figure 9.5 Battery module assembly

Each cell has to be provided with an external fuse for protection of the central store against cell failure during a short-circuit mode. A fuse has also to be provided at any central store terminal to protect the cell fuses from an external system fault. Power contactors and manual circuit breakers have to be used to disconnect the central store from the DC bus for the holding mode or for maintenance.

A simplified electrical schematic of a central store is shown in Figure 9.6.

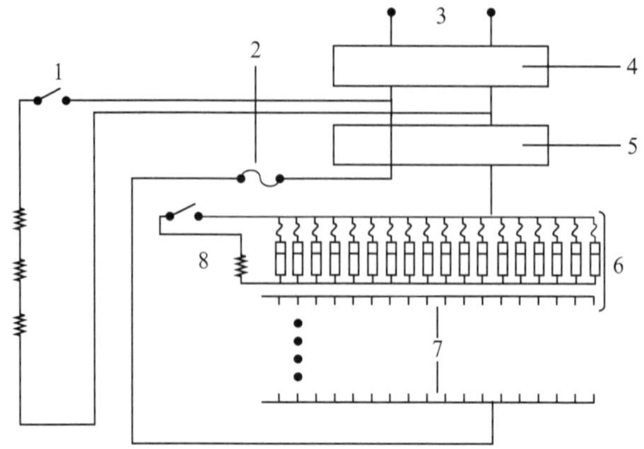

1. Reserve heater circuit	5. Contactors
2. Bus fuse	6. 18 cells and fuses in parallel per module
3. To DC bus	7. Total of 78 modules inseries
4. Circuit breaker	8. Bypass heater (one per module)

Figure 9.6 Simplified electrical schematic diagram of a battery

9.5 Thermal regime

Lead-acid cells do not require any special temperature regime, but if advanced batteries – such as Na-S or Li-S cells – are involved, maintenance of the temperature regime becomes important. The thermal control equipment comprises electric heaters within the central store's core, and heat exchange panels that form the interior ceiling of the CS enclosure. Natural convection currents within the CS could be promoted by an inclination of the interlayer separators and enclosure ceiling, as shown in Figure 9.7. The strip heaters include module bypass and reserve units providing redundant capability for maintaining a minimal cell temperature under all conditions.

Owing to the cells' internal resistance, there is significant waste heat rejection during the battery charge and discharge modes. Internal heat dissipation increases over the life of a cell due to increased internal resistance with age.

Dissipation is also greater at the end of discharge owing to entropic heating and to reduced cell voltage, and, hence, greater battery current for the same delivered power.

Battery thermal losses are approximately linear with cell/ambient temperature difference and, despite the significance of these losses, they are comparable to the waste heat rejection rates of the cell during both charge and discharge regimes.

During charge, the waste heat rejection exceeds the thermal loss and the battery temperature rises without heater operation. Normally, a keeping regime will follow the charge period. Operation of the heaters during this idle period can then complete the recovery of battery temperature to the required level.

The internal heating rates are substantially greater on charge than on discharge. This is due to the higher charging rates required for recovery from extended discharge and is only slightly compensated by entropy effects within the cells.

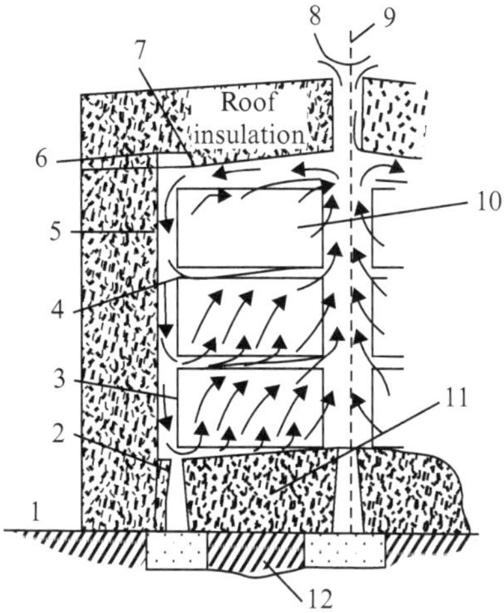

1. Ground level	7. Cooled ceiling
2. Piers	8. Cooling air vent
3. Side baffles	9. Battery midplane
4. Interlayer separators (with heaters)	10. Module layer
5. Wall insulation	11. Floor insulation
6. Cooling air duct	12. Foundations

Figure 9.7 Cross-section view of a battery

So special heater operation, and therefore energy consumption, is needed during the discharge regime. To reduce the need for heater operation on discharge, it would be reasonable to use heat accumulated in the core of the central store during the charge regime, allowing the battery to cool down only to a certain temperature, say 300 °C for Na-S cells, for example. This certainly restricts the permissible duration of the storage-keeping regime, but for daily regulation storage no additional heat is needed.

It should be noted that thermal losses do not affect the system's peak power requirement. Once the battery is fully charged, the available system power is more than enough for heater operation. Also the thermal losses do not impact on the required storage energy capacity. At the beginning of storage life, the redundant cells provide the necessary energy capacity to satisfy the heater requirements for extended discharge at the rated energy output. At the end of the cell's life, no heater operation is required at all for extended discharge.

9.6 The power extraction system

Any secondary battery, or cell, uses direct current. Since all power systems are AC, a power transformation system for this kind of storage has to convert DC to AC. The most efficient way of doing this is by the use of a thyristor-based rectifier and inverter. Since, during their operation, both the rectifier and the inverter consume reactive power, special reactive power compensation equipment is needed. The AC/DC rectifier/inverter has to be coupled to a power system with the help of an

AC transformer since the system voltage is usually much higher than the storage working voltage. Hence the PTS for battery storage has (at least) to comprise AC/DC rectifier/inverter, reactive power compensation equipment and AC transformer.

There are two different types of AC/DC convertor design:

- line-commutated and
- force-commutated.

A line-commutated convertor has a lower capital cost and, in addition, no special precautions are needed to protect the storage plant from faults on the AC system.

The main advantage of the more expensive force-commutated convertor is allowing the storage at, for example, an 11/33 kV substation to supply consumers on the 11 kV side even after disconnection on the 33 kV side. If not specially designed, the battery alone would hardly be able to support the substation load, but it is extremely valuable for power system stability maintenance as well as for so-called uninterruptible power system designs.

Battery storage is shown schematically in Figure 9.8. Thyristor-based PTS will be considered in more detail in Chapter 10.

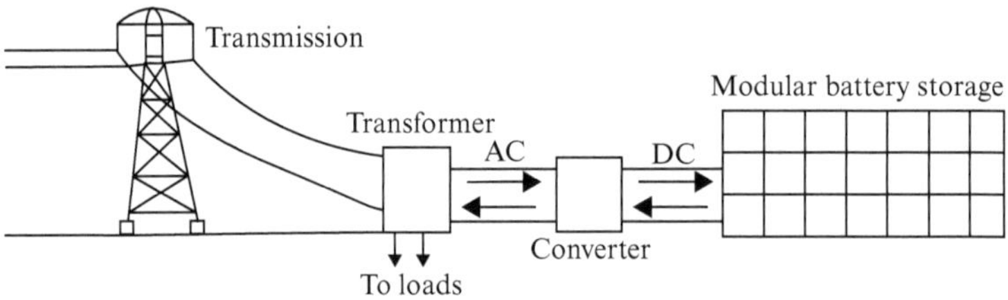

Figure 9.8 Energy storage in electric batteries

Chapter 10
Capacitor bank storage

10.1 Theoretical background

Energy can also be stored in the form of an electrostatic field. Let us consider an electrical capacitor, that is, a device that can collect electric charge which is establishing an electric field and hence storing energy. The capacitance C of a capacitor is defined by the amount of charge q it can take up and store per unit of voltage:

$$C = \frac{q}{V_c}$$

where V_c is the voltage of the capacitor.

As an example, let us take a plate capacitor with plate area A and distance d between its plates, as shown in Figure 10.1. From the definition of capacitance it follows that the capacitance of the capacitor is

$$C = \frac{\mathrm{kA}}{d}$$

The so-called dielectric materials can be polarised and, if used as a medium, the total charge stored in a capacitor using such a material will be increased.

The properties of this medium may be described by the constant k called the permittivity, which is measured in F/m since both A and d have metres as the basic unit. The electric field E is homogeneous inside the plate capacitor only when the distance between the plates is small. In order to keep the charges at the plates divided, and thereby to maintain the electrostatic field, the dielectric medium must have a low electronic conductivity; so we are looking for dielectric materials with high permittivity.

Let us consider a series RC circuit that is connected to a constant voltage source V as shown in Figure 10.2. When the switch is closed a transient process that can be described by Kirchhoff's circuit equation begins:

$$V - i(t)R - V_c = 0$$

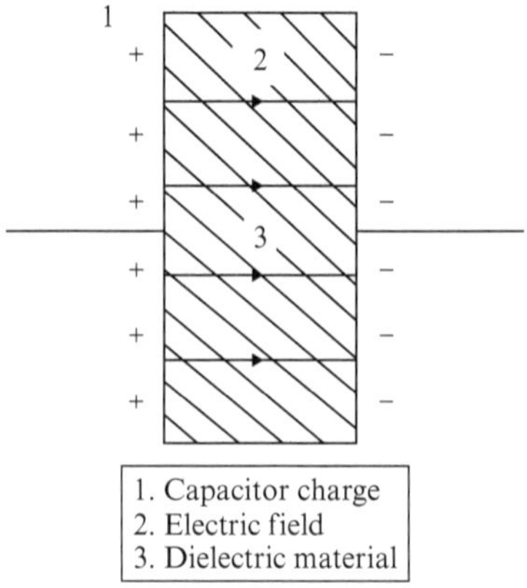

Figure 10.1 Parallel plate capacitor

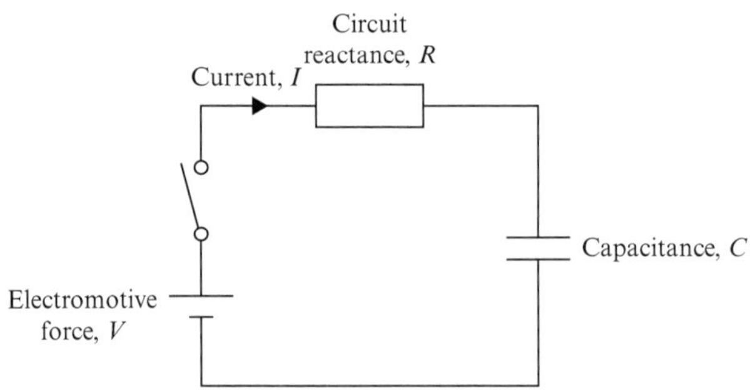

Figure 10.2 Series RC circuit

According to the basic definitions voltage and charge are described by $V_C = qlC$ and $q = i(t)dt$; then Kirchhoff's circuit equation is given by

$$\frac{V - i(t)R - i(t)dt}{C} = 0$$

After differentiation and rearrangement, Kirchhoff's circuit equation can be rewritten as

$$\frac{-Rdi(t)}{dt} - \frac{i(t)}{C} = 0$$

or

$$\int \frac{di(t)}{i(t)} = -\int \frac{dt}{RC}$$

The solution of this equation is shown in Figure 10.3 and may be given as follows:

$$i(t) = I_0^{(-t/RC)}$$

where $i(t)$ = circuit current at a given time t
I_0 = circuit current at a zero time
R = total circuit resistance, comprising capacitor inner resistance R_i, connecting lines resistance R_c and source resistance R_s: $R = R_i + R_c + R_s$.

The current at $t = 0$ is limited by the total resistance R:

$$I_0 = \frac{V}{R}$$

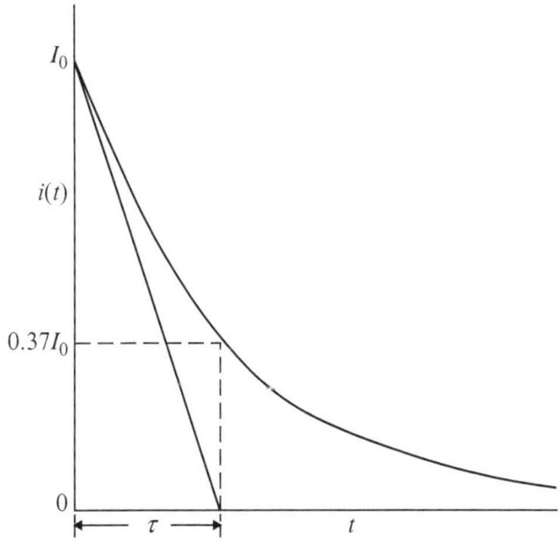

Figure 10.3 Transient response of a series RC circuit

According to the definition the time constant τ is the time taken to end the process if the rate of change was constant. The time constant can be calculated by differentiating the expression for current, taking into account that the value of $i(t)$ at $t = 0$ is I_0. Hence the time constant is the capacitance times the total circuit resistance:

$$\tau = RC$$

It follows from the solution of Kirchhoff's circuit equation that the current at $t = \tau$ is $0{\cdot}37I_0$. The time it takes to charge a capacitor equals 5 τ since, by that time, the whole charging process has effectively been completed. The time constant for discharging is exactly the same as for charging, but the current, being equal to zero at the beginning of the discharge regime, becomes equal to I_0 at the end of the process.

Let us calculate the amount of energy stored in the capacitor. Since voltage is defined as work or energy W per charge unit:

$$V = \frac{dW}{dq} \quad \text{or} \quad dW = Vdq$$

Kirchhoff's voltage equation can be transformed to an energy equation by multiplying each term by charge. The energy equation will therefore be as follows:

$$\frac{Vdq - i(t)Rdq - qdq}{C} = 0$$

or

$$dW_t = dW_r + dW_c$$

It is clear from the energy equation that, of the total amount of energy dW_t consumed by the power source, a certain part dW_r is wasted as heat in circuit resistance and a certain part dW_c is stored in the capacitor. The increment of stored energy is given by

$$dW_c = \frac{qdq}{C}$$

So the energy stored in the capacitor may be calculated as

$$W_c = \frac{0.5\, q^2}{C}$$

or, since by definition $q = CV_c$, this solution can be rewritten as

$$W_c = \frac{CV_c^2}{2}$$

or

$$W_c = \frac{qV_c}{2}$$

The total amount of energy W_s supplied to the capacitor by the extremal source during a charge period is qV_S. So it is not difficult to calculate the charging efficiency ξ_S, which will be given as

$$\xi_s = \frac{W_c}{W_s} = \frac{0.5\, qV_c}{(qV_s)}$$

Since at the end of the charge period $V_c = V_S$, this efficiency is exactly 50%, which is not particularly impressive.

The only possible way to increase the charge efficiency of capacitor bank storage is to control the power source voltage V_s so that the charging current is constant.

By definition the volume energy density is given as follows:

$$w = \frac{dW}{d(\text{Vol})}$$

In the case of a homogeneous field w is constant and therefore

$$w = \frac{W}{(\text{Vol})} = \frac{0.5\, CV_c^2}{(dA)}$$

At a given voltage V_c, the energy contained in the electric field is proportional to the capacitance, which again is dependent on the medium in which the field exists.

Using the relationship between the electrostatic field and the voltage, $E = V/d$, as well as formulae for the parallel plate capacitor's capacitance

$$C = \frac{\kappa A}{d}$$

we get

$$w = \frac{0.5\ \kappa A E^2 d^2}{(d^2 A)} = 0.5 \kappa E^2$$

10.2 Capacitor storage media

The quantity of energy stored is dependent on the ability of the material to polarise when placed in the electric field between the plates of the capacitor. Most insulators have a relative permittivity (k_r) of 1–10 compared with that of vacuum or dry air (k_0), the relationship being

$$k = k_0 k_r$$

All materials that comprise dipoles may be polarised and they have a relative permeability, also called the dielectric constant k_r, greater than unity. Teflon, for instance, with a k_r value of approximately 2, will double the energy content when placed between the plates of a capacitor. Some compounds, such as titanates, have k_r values up to 15 000, and they are therefore often used as dielectric materials in capacitors, especially where small size and large capacitance are required. It should be noted that the farad is a rather large unit and, in practice, capacitance is measured in units of picofarads (1 pF $= 10^{-12}$ F) or microfarads (1 mF $= 10^{-6}$ F). The whole Earth as a spherical capacitor has a capacitance of much less than 1 mF. In order to calculate how much energy the radial field of the Earth contains, if the Earth were charged to thousands of volts relative to deep outer space, the energy stored in its radial electrostatic field would be of the order of magnitude of 1 J.

Calculations show that ordinary capacitors can store only very limited amounts of energy because of their small capacitance. For example, let us take a 10^7 V/m field in a good insulator with $k = 10^{-11}$ F/m. For these conditions the volume energy density will be

$$w = 0.5\ kE^2 = 5 \times 10^2\ \mathrm{J/m^3}$$

This, of course, is a low value compared with chemical storage.

The development of solid ionic conductors in recent years provides the prospect of capacitance in the range of 1 F in a volume of only 1 cm^3, many orders of magnitude more than the best dielectric materials.

Most capacitors have extremely small internal resistance R_i. This means that the power density can be quite high when a capacitor is short-circuited compared with chemical storage with high internal resistance. Therefore capacitor bank energy storage will find valuable applications where a source of very high power is required. Where electrochemical processes are also involved, such storage media will be of importance in future in vehicle regenerative braking systems, for example, where the energy requirement is moderate but the power requirement is essential.

10.3 Power extraction

The power transformation (extraction) system for capacitor bank storage is practically the same as for chemical storage. It uses a thyristor-based AC/DC convenor and all the relevant devices: AC transformer, reactive power sources, etc. A circuit schematic for a PTS is shown in Figure 10.4. The main requirement for a capacitor bank energy storage PTS is the necessity to change the polarity of the central store when changing working modes from charge to discharge. This requirement doubles the PTS size, and therefore cost, compared with that for a magnetic energy storage system (which will be considered in Chapter 11).

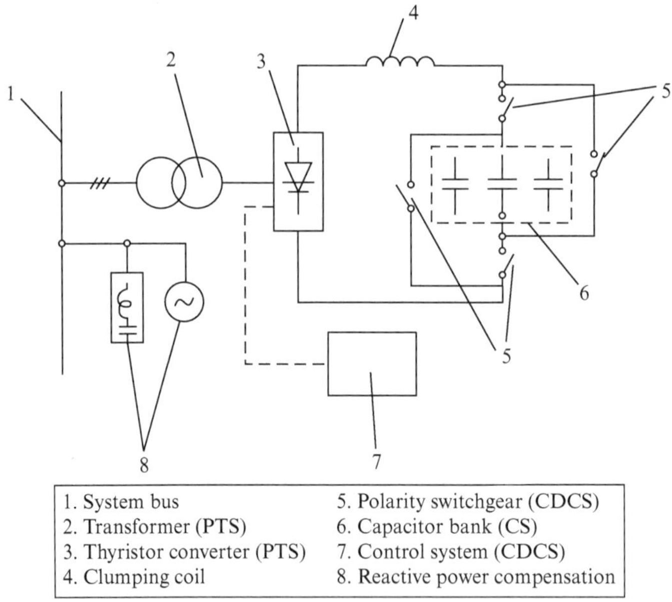

Figure 10.4 Circuit diagram of a capacitor bank energy storage system

Chapter 11

Superconducting magnetic energy storage

11.1 Basic principles

Completely novel, based on the development of superconductors, is the possibility of storing significant quantities of energy in magnetic fields.

Let us consider a series *RL* circuit – a simplified diagram of superconducting magnetic energy storage (SMES) as shown in Figure 11.1. Here *R* is the total resistance of the electrical circuit between the source of the power supply of voltage *V* and the magnetic coil of self-inductance *L*. The total resistance comprises the internal resistance of the source and resistance of the coil. When a coil is connected to a constant voltage source, the electric current varies with time: it is zero at $t = 0$ and it stabilises at I_{max} when the magnetic field has been built up. If the switch in Figure 11.1 is closed, a transient storage process will start and electrical current $i(t)$ through the circuit will rise during charging and fall during discharging. The transient process equation may be written as follows:

$$i(t) = \frac{(V + e_s)}{R}$$

where e_s is the induced electromotive force (EMF)

$$e_s = \frac{-L\,di(t)}{dt}$$

Since $i(t) = 0$ while $t = 0$, the solution of the transient process equation will be given as follows:

$$i(t) = I_{max}\{1 - e^{(-Rt/L)}\}$$

and is shown in Figure 11.2. Here $I_{max} = V/R$. The time constant is L/R and its form is shown in Figure 11.2.

Multiplying each term of the transient process equation by $i(t)dt$, we can get the energy equation:

$$\frac{Vi(t)dt - i^2(t)Rdt - Ldi(t)}{dt\;i(t)dt} = 0$$

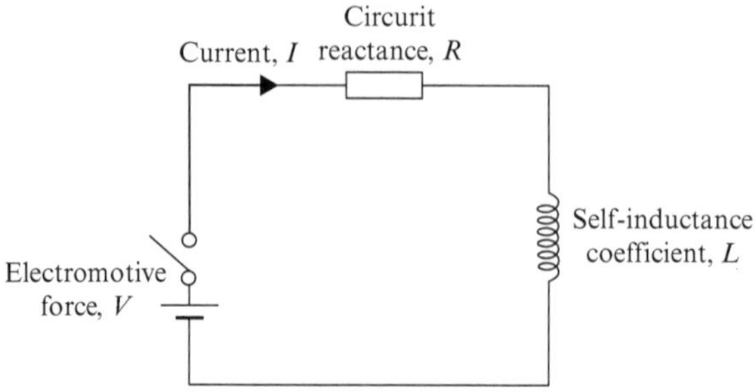

Figure 11.1 Series RL circuit

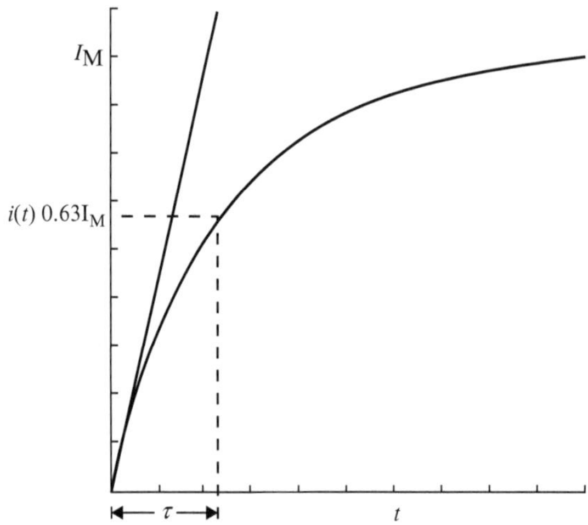

Figure 11.2 Time variation of current in an RL circuit

or

$$dW_s = dW_r + dW_m$$

The energy supplied from the external source, dW_s, divides into two parts, one of which is the energy content of the magnetic field (dW_m) and the other is the energy losses. It takes energy to build up the magnetic field and that energy can be released again during discharge as an electric current in a circuit containing load.

The energy stored in the magnetic field, dW_m, is given by

$$dW_m = Li(t)di(t)$$

Since L is a constant, it follows that

$$W_m = L\int_0^{I\max} i(t)\ di(t) = 0.5\ LI_{\max}^2$$

The energy content in an electromagnetic field is determined by the current I through the N turns of the coil of the magnet. The product NI is called the magnetomotive force. Using another expression for induced EMF in the coil, $e_s = -Nd\phi dt$

$$W_m = \int_0^{\phi} Ni(t)\ \phi = \int_0^{B} lAH\ dB = vol \int_0^{B} H\ dB$$

Since there is an approximately linear relationship between H and B, the solution will be given by

$$W_m = 0.5\ \mathrm{Vol}\ \mu H^2$$

or

$$W_m = \frac{0.5\ \mathrm{vol}\ B^2}{\mu}$$

where ϕ = magnetic flux
B = induction
l = length of the magnetic field
A = area of the magnetic field
N = number of turns of the coil
μ = permeability
L = self-inductance

Therefore, at any moment of time, the stored energy is given by half the product of the inductance and the square of the current, or by half the product of the flux density and the magnetic field integrated over the whole volume in which the field is significant. The volume energy density can be obtained from these equations as follows:

$$w = 0.\ 5\ \mu H^2 = 0.5HB = \frac{0.5B^2}{\mu}$$

The energy density that can be achieved in a magnetic field of, for example, $B = 2T$, is approximately $2 \times 10^6\ \mathrm{J/m^3}$, which is an order of magnitude or more greater than that attainable in an electrostatic field. However, it is still not too impressive in comparison with chemical batteries.

A superconducting coil can be connected to a constant DC power supply. As the current of the coil (which is pure inductance with inner resistance equal to 0) increases, the magnetic field also increases and wholly electrical energy is stored in the magnetic field. Once I_{max} is reached, the voltage across the coil terminals is reduced to 0. At this stage, the coil is fully charged and the energy can be stored as long as is desired. In contrast, a conventional coil made of copper windings with a particular resistance would require continuous power input to keep the current flowing.

The idea of storing electricity in a very large superconducting magnet is, at first sight, very attractive and might be arranged as shown in Figure 11.3. Having negligible losses, such a storage system would have a very high efficiency, could feed straight into the electrical system and, theoretically, could be built on a very large scale.

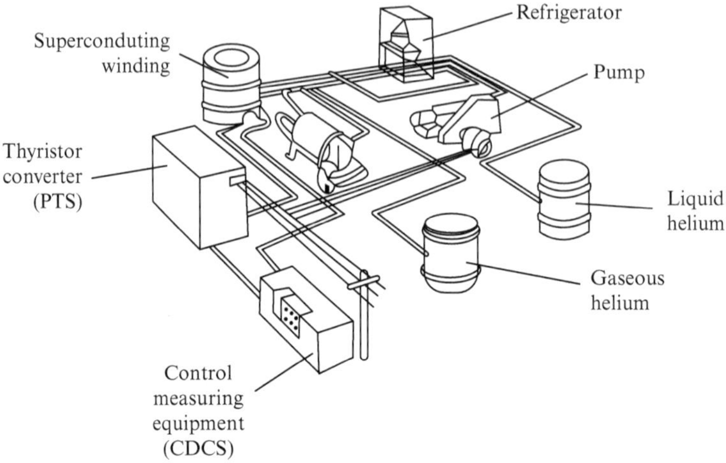

Figure 11.3 Experimental SMES composition

The following specific problems need to be resolved before SMES can be in widespread use for storage in power systems:

- compensation for stray fields;
- effects of electromagnetic forces on conductors and support elements; and
- protection against sudden appearance of normally conducting zones.

Let us now consider these in more detail.

11.2 Superconducting coils

Superconductors are able to carry very high current in the presence of high magnetic fields at low temperatures with zero resistance to the steady flow of electrical current. Unless their critical values – temperature T_c, induction B_c and current density S_c – are exceeded, superconductors show no resistance and thus are able to carry high direct current without any losses. The high current density allows a device to be considerably more compact compared with conventional devices designed for the same application. All these factors suggest that superconductors will find useful applications in power systems.

Technical superconductors are normally made of NbTi or Nb_3Sn multifilaments embedded in a copper or aluminium stabilisation matrix, as shown schematically in Figure 11.4. At present, Nb_3Ti superconductors are mainly used

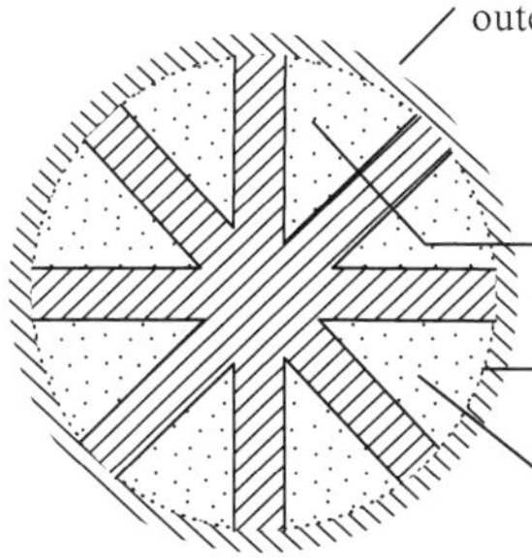

Figure 11.4 Superconductive wire seen in cross-sectional view

for reasons of ease of manufacture. The critical values of $Nb_{47\%}.Ti$ range from 0 up to

$$T_c(B = 0, S = 0) = 9.2\ \text{K}, B_c(T = 0,\ \ S = 0) = 15\text{T}$$

and

$$S_c(B = 0, T = 0) = 10^4\ \text{A/mm}^2$$

Current density S_c decreases with increasing values of B and T, for example

$$S_c(B = 5\ \text{T}, T = 4\ 2\ \text{K}) = 2300\ \text{A/mm}^2$$

If the superconductor becomes normally conducting, the current transfers to the stabilisation matrix, thus avoiding destruction of the superconductor by overheating.

Since ferromagnetic materials do not apply to inductions higher than $3T$, coil arrangements used for superconductor magnetic energy storage are usually placed in such media as air or vacuum with $\mu = 1$. In order to obtain high values of W_m with a given coil current I, which is limited by the applied superconductor, the total self-inductance L of the storage has to be made as high as possible by choosing a suitable coil geometry.

There are three concepts of superconducting magnetic energy storage (SMES) design:

- circular-shape single solenoid;
- series connection of coaxial solenoids; and
- circular, oval or D-shape torus comprising series-connected single coils.

Solenoids are characterised by the length/diameter (aspect) ratio P. Long solenoids with $P >> 1$ store magnetic energy mainly inside the coil, flat solenoids with $P < 1$ in the exterior space of the coil. Flat solenoids make more efficient use of a given quantity of superconductor. The electromagnetic forces of a solenoid, as shown in Figure 11.5, are work-expansive in radial and work-compressive in axial directions.

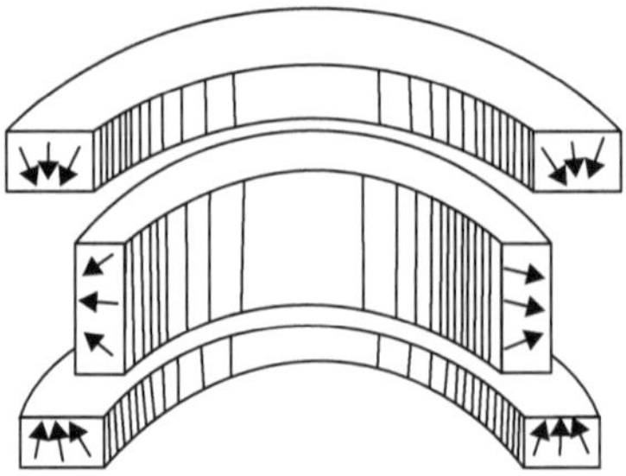

Figure 11.5 Winding stress diagram

Toroidal storages, composed of single coils, have a minimal external field, but use about double the amount of superconductor. The electromagnetic forces are dependent on the circumference coordinates of the single coils, which causes a resulting radial force towards the centre of the torus.

Some values of superconducting coils for different SMESs are given in Table 11.1. All the data are for a working temperature of 1.8 K, a current of 757 kA and a magnetic field of 6 T.

Table 11.1 Parameters of coils

Energy capacity	**3.6×10^{13} J**	**3.6×10^{12} J**	**3.6×10^{11} J**
Outer radius (m)	100.5	74.5	34.6
Inner radius (m)	146.2	68.0	31.6
Height (m)	16.2	7.5	3.5
Number of turns	2675	1240	576
Inductivity (Hn)	2920	292	29.2
Force (N)	6.85×10^{11}	1.48×10^{11}	0.319×10^{11}

All SMES projects for power system application suggest coils with a high magnetic field. A field of 5 T, for example, is equivalent to a pressure of 100 atm. The design of large coils is therefore dominated by the enormous electromagnetic forces tending to burst them and, for solenoids, also to crush them axially. The cost of a self-supporting structure to contain these forces would almost certainly make an SMES extremely expensive, so all proposals for large SMES systems suggest placing the windings in circular tunnels cut in bedrock. If the installation is deep enough, the weight of the overbearing material ensures that the forces in the rock remain predominantly compressive, and therefore compensate the SMES electromagnetic forces. Since the thickness of the winding and its enclosing vacuum and helium jackets will be small, only a relatively small amount of rock needs to be excavated.

To store the energy applicable in the power system, a very large coil of approximately 100 m diameter is needed and, in general, the minimum amount of material for a given energy will be required.

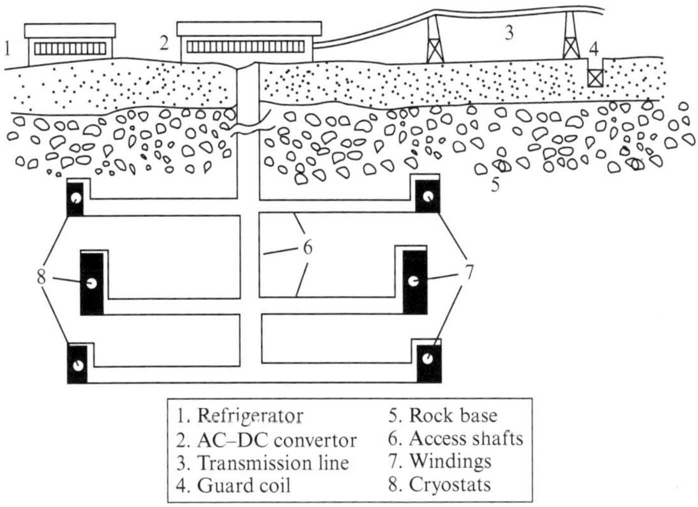

Figure 11.6 Cross-sectional view of a conceptual 10 000 MWh underground superconducting energy store with the main winding divided into three sections

SMESs in the range of 10^{13} J should be designed according to the concept of series-connected coaxial solenoids. SMESs up to 10^{9} J are preferably designed to be circular-shape single solenoids or circular-shape toruses, which are more suitable.

A schematic arrangement for a 10 000 MWh or 36×10^{13} J SMES, based on work done at the University of Wisconsin, is shown in Figure 11.6. Dividing the winding into three, in a stepped arrangement as shown in Figure 11.5, helps to reduce the field and the axial compressive forces at the ends of the winding, and provides additional rock ledges for their support. Some typical values for the dimensions and operating parameters of this type of store are given in Table 11.2.

Table 11.2 Typical parameters for a large superconducting magnetic energy storage system

Total stored energy	10 000–13 000 MWh
Available energy	9000–10 000 MWh
Discharge time	5–12 h
Maximum power	1000–2500 MW
Maximum current	50–300 kA
Maximum field (occurs at the conductor)	4–6 T
Mean diameter of winding	300 m
Total height of windings	80–100 m
Mean depth below surface	300–400 m
Efficiency (assuming one complete charge/discharge cycle per day)	85–90%
Convertor losses	2% of mean power rating
Refrigerator drive power	20–30 MW

A sound and stable geological formation is needed, but the scale is not large by mining standards.

11.3 Cryogenic systems

The cryogenic system of an SMES device comprises a refrigerator, where a coolant has to be prepared, and a cryostat-storage vessel where a superconducting coil is placed to be refrigerated and thermally isolated from the environment.

The cooling system is usually based on liquid helium as coolant, either in a helium bath or by forced circulation. This removes all heat entering the cryostat and thus ensures the superconductor temperature does not exceed its critical value anywhere. With regard to the limited efficiency of the refrigerating plant, the heat input through supply leads, mechanical supports and radiation should be as low as possible. It can be minimised by providing intermediate cooling, so-called thermal shields. Thermal effects caused by the SMES electrical current are normally small compared with the heat transfer mentioned above.

The refrigerator consumes electrical energy, thus decreasing the SMES efficiency. The refrigerator's power consumption P_{ref} ranges are approximately 300–1000 times the cooling power (thermal output); the refrigerator's cycle efficiency ξ_{cyc} is given by

$$\xi_{cyc} = \frac{E_s}{(E_s + t_{cyc}P_{ref})}$$

where E_s = stored energy
t_{cyc} = cycle duration.

A value for cycle efficiency is essential to calculate the total efficiency of an SMES system.

Coil refrigeration and thermal insulation are extremely difficult technical problems, since low temperatures (around 1.8 K) are needed to enhance the superconductor's current-carrying ability and take advantage of the improved heat transfer provided by superfluid helium II.

Circular-shape vessels containing the superconducting coil must be placed in a vacuum stratum to prevent any heat transfer from outside to the cooled part. At the same time, mechanical stresses have to be transmitted from the coil to the supporting rocks through special struts with as low a thermal conductance as possible. It has been proposed to use a corrugated arrangement of both conductor and vessel walls; such an arrangement reduces the tensile forces in the conductor and allows for movement caused by magnetic pressure and thermal contraction. In addition, it becomes possible to use thin walls for the cryostat, which usually comprises a number of thermal shields, as shown in Figure 11.7. Each of these is refrigerated to a certain temperature in order to reduce the heat load at the lowest temperature, 1.8 K, as much as possible. One or more points on each of the supports must also be refrigerated (see Figures 11.7 and 11.8).

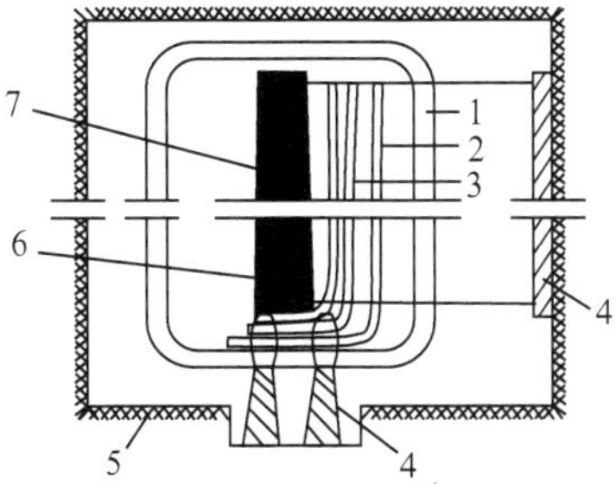

1. 80 K temperature screen	5. Shaft
2. 20 K temperature screen	6. 1.8 K liquid helium
3. 4.2 K temperature screen	7. Superconductive winding
4. Cryostat support	

Figure 11.7 Cryostat in cross-section

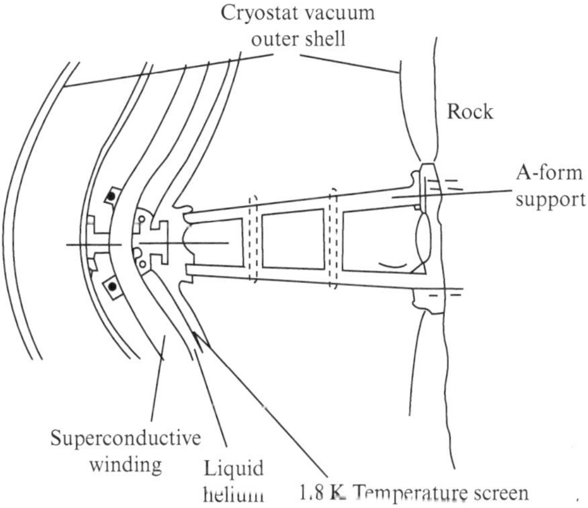

Figure 11.8 Winding support seen in cross-section

11.4 Power extraction

A superconducting coil is, in fact, a source of variable direct current. To couple this source to a constant voltage AC power system, a special power transformation system is required. Rectifying and inverting systems of this type are already in use for coupling DC connection lines to the AC power system. By adjusting the thyristor's delay angle, smooth and rapid change of charge or discharge rate, and even a quick change of regime, becomes possible within one cycle of the power system frequency.

A typical SMES configuration comprises two 6-pulse thyristor Greatz bridges series-connected to the superconducting coil on the DC part of the bridge and coupled through an AC transformer to a power system on the AC side of the bridge, as shown in Figure 11.9.

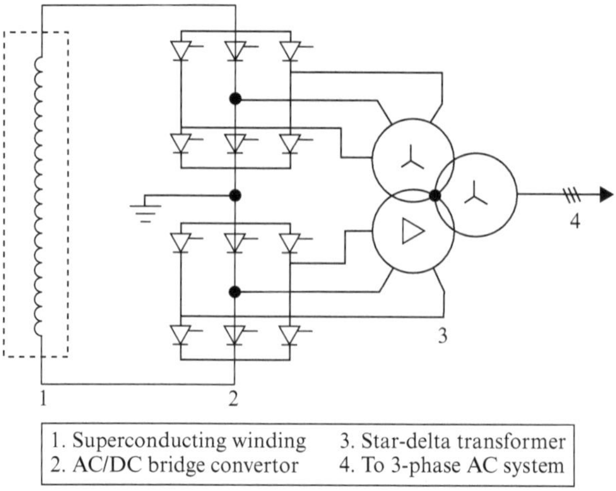

Figure 11.9 Basic circuit for superconducting magnetic energy storage

Since any convertor is a considerable consumer of reactive power, its compensator is an essential part of this type of PTS for SMES.

An attractive feature of such a PTS is its high efficiency: losses attributable to the solid-state bridge conversion (AC/DC and DC/AC) are estimated to be between 3% and 8% of the total stored energy. At a given coil current (I_d), the proper delay angle is calculated for the required active and reactive convertor power, P and Q; this calculation is typically performed at 30 ms intervals.

When the phase delay angle p is less than 90°, the bridge operates in rectifier mode and acts as a load of $S = P + jQ$ for the AC power system. Alternatively, if p is set above 90°, the average convertor voltage becomes negative, the active power P changes sign and the bridge becomes a kind of power source for the AC system. This is caused by the fact that only unidirectional flow of current is possible through the convertor. Therefore, with $p > 90°$, the 3-phase bridge operates in inverter mode. The rectifier/inverter behaviour of the convertor can easily be explained by its circular *PQ* diagram that describes the operating conditions in the voltampere plane.

Proper control of the delay angles of these bridges makes it possible to control the rate of active-power consumption/generation and reactive-power consumption independently, rapidly and smoothly at the bus where an SMES is placed, and therefore all thyristor convertor-based storages can be applied to power system stabilisation.

Line-commutated convertor bridges have a limited delay angle of $p_{max} = 140°$, thus avoiding the risk of commutation failure. It is clear that this limit substantially reduces the operational range of the convertor bridge when working in the inverter mode. An additional limitation may arise from the harmonic impact of AC bus currents. This limitation is difficult to quantify in terms of the thyristor delay angle because the total harmonic distortion (THD) of the AC bus voltage and current depends on the AC network parameters and scheme.

Simultaneous control of both active and reactive power can only be achieved using thyristor Greatz bridges over a limited range, and this, of course, limits the effectiveness of SMES applications for power system stabilisation. On the other hand, recent developments of the gate turn-off thyristor (GTO) allow us to design a convertor that has the ability to control power within a circular range containing four quadrants in the active and reactive power domain. It is in this way that it is likely that the effectiveness of using SMES as a stabiliser in power systems will be improved.

11.5 Environmental and safety problems

An obvious problem associated with solenoidal magnetic coils is the compensation of external stray fields. The field falls off approximately as the inverse cube of the distance from the centre of the coil, but one must go 1 km from the centre of the SMES shown in Figure 11.6 before the field has dropped to a few millitesla, and even this is not enough since, for example, the Earth's field is approximately one-twentieth of a millitesla. Since sufficient reduction cannot be obtained by reasonable distance and depth alone, a guard coil with a magnetic moment equal – but in the opposite sense – to that of the SMES central store has to be added to the SMES system, as indicated in Figures 11.10 and 11.6. By using a much larger radius, fewer ampere-turns are required and the decrease in stored energy is a few per cent

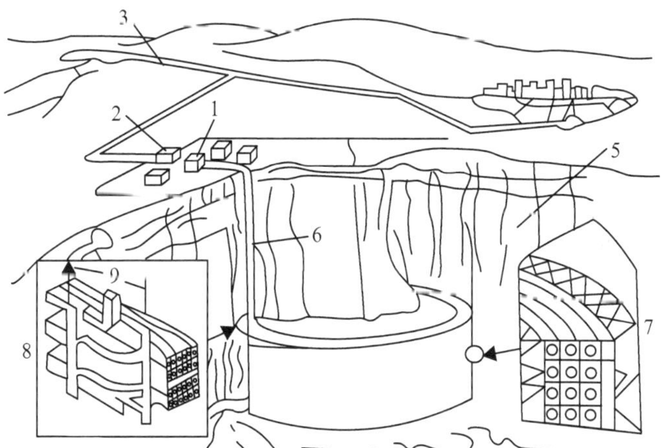

1. Refrigerator	6. Access shafts
2. Converter	7. Superconducting windings
3. Transmission line	8. Cryostat
4. Guard coil (not shown)	9. Liquid helium
5. Rock	

Figure 11.10 General plan of the concept of a 10 000 MWh underground superconducting energy store with the main winding, cf. Figure 11.6

only. The stray field then falls off roughly as the inverse fourth power of the distance and at 1 km would be down to just a few tenths of a millitesla. The exact level to which the outer field must be reduced in order to avoid negative effects on people, power transmission lines, aircraft and bird navigation, nearby ferrous metals and so on has not been investigated yet, but it does not seem that the problem differs seriously from that of high-voltage transmission lines.

As any other superconducting device, operational SMES always runs the risk of the sudden appearance of normal conducting zones and their expansion, so-called 'quenches'. To reduce the probability of quenches to a minimum, it is recommended that well-stabilised superconductors are used with a short distance between the superconductors and cooling tubes. Normal conducting zones are then suppressed by removing all ohmic heat by thermal conduction and by coolant without increasing the temperature of the adjacent sections above their critical values. Single quenches cannot, however, be excluded totally, and an external protection system is therefore needed. As materials at low temperatures have very small heat capacities, loss of superconductivity leading to ohmic heat release may cause a dangerous rise in temperature and destroy, if not blow up, the superconducting coil. The detection and handling of quenches is an essential part of any SMES system.

A superconducting coil is an inductance without resistance, which, in normal operation, is equal to an EMF only. Any additional voltage rise is likely to be caused by normal conducting zones and therefore indicates a quench. Quench detection systems, therefore, should be based on measurement of the single coil voltage for a torus, or coil section voltage for a solenoid, and of the time derivation of the current. If the measured value of V differs from the EMF calculated on the basis of the first derivative of current with respect to the time measurement, a quench has taken place and protective measures must be started.

However, eddy currents can cause parasitic voltages by magnetic coupling and, in this way, can simulate a quench. Therefore, it is better to compare the voltage difference with a particular threshold voltage which allows consideration of the uncertainties in measurement and the effect of eddy currents.

A thermal supervision system that controls both coil temperature and magnetic induction at the maximum points should also be added to the electrical quench detection system (Figure 11.11). This will enable control of storage current to the maximum permissible at any time, and avoidance of quenches caused by exceeding the superconductor's critical values (if the actual values are known).

After the detection of a quench, the damaged coil must be protected against local overheating by discharging it promptly; otherwise the zone with normal resistance would expand quickly along and across the coil windings causing ohmic heat release, which can seriously affect the cooling circuit.

There are two methods to get the coil quickly discharged: (i) discharge of all central store coils and (ii) discharge of the affected coil only.

De-excitation according to the first method can be done either by discharging back the stored energy into the power system or by converting it into heat through the resistance situated outside the refrigerated storage vessel. Since the total inductance of the central store is high, prompt discharge requires high voltages,

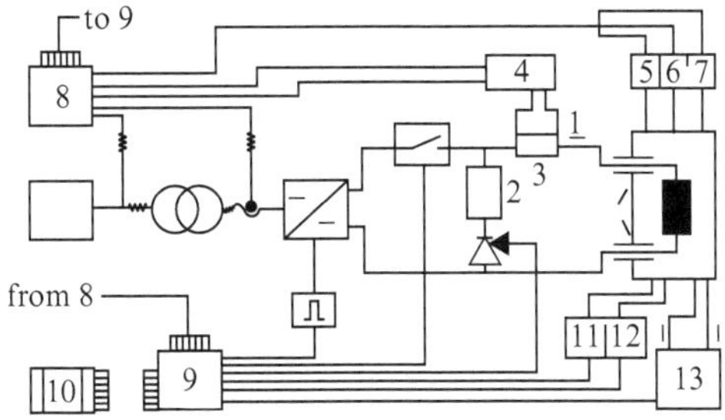

1. Cryostat and supply leads	8. Measured values processing
2. Discharge resistance	9. Master unit
3. Shunt	10. Control panel
4. Current-measuring device	11. Heaters trigger control
5. Magnetic induction measuring device	12. Bypass thyristors trigger control
6. Helium temperature-measuring device	13. Liquid helium-refrigerating plant
7. Coil voltage-measuring device	

Figure 11.11 Principle arrangement of a control and protection system for SMES (network control not included)

which are strictly limited by the power transformation equipment – AC transformer and AC/DC convertor – and by the supply leads into the refrigerated storage vessel, a cryostat.

An alternative method that guarantees very fast de-excitation and a homogeneous distribution of released heat by means of internal ohmic heaters is not recommended, since all the stored energy is released inside the cryostat. The temperature of the coil does not exceed the limiting values, but it would be necessary to evaporate the helium before it expands and blows the cooling tube to pieces. The consequence is a long cooling down period during which the SMES stands idle.

The coils of a toroidal central store can be discharged separately using bypasses like psn-diodes or thyristors. The superconducting dipole and quadrupole magnets of the accelerator plants HERA and CERN are bypassed by diodes. Since an SMES coil has considerably higher coil reactance voltages, thyristors are needed to bypass it.

Bypassing keeps the central store functioning and enables the damaged coil to be recooled. In torus-shaped coils, each bypassed coil decreases the total inductance of the central store, which leads to a current increase, since the magnetic field cannot be changed immediately. If the current is going to exceed the limit, which may be either the actual critical current of the superconducting winding or the maximum permissible current of the supply leads and the AC/DC convertor, no additional coil may be bypassed, even if it is affected by a quench itself.

As the current commutates more rapidly to the bypass thyristor at higher coil voltages that are caused by the normal zone's resistance, it would be reasonable to fire the coil's heaters, thus creating a normal homogeneously conducting coil, to avoid hot spots in the starting region of the quench. A residual current keeps flowing in the coil, increasing as the coil's temperature decreases and thus preventing it from becoming superconducting. The central store must be discharged in any case, whether or not the coil is bypassed. However, bypassing enables the protection system to save almost all the stored energy.

Bypassing is also applicable to solenoid-shaped central stores, if the coil is divided into a certain number of sections. The magnetic connection between the sections, being higher than between the single coils of a torus, improves the current commutation from the coil section to the bypass.

To ensure short outage time and full energy saving in the case of failure, SMES protection systems should be designed to be selective. By permanent control of current and coil voltage, it is possible to detect a quench immediately. By measuring the temperature and obtaining the magnetic induction, this fixes the actual permissible current, and, by keeping a secure difference between it and the superconductor's critical values, reduces the probability of quenches. In the case of a quench, the protection system first fires the bypass thyristor and the heater of the affected coil and switches the AC/DC convertor to inverter mode, thus saving the stored energy or discharging it into the power system. If the power system is not able to absorb the energy, the SMES should be separated from it, and the stored energy will be exhausted through the discharge resistance. If more coils become normally conducting, they are bypassed unless the current exceeds its permissible value. In that case, all the heaters are fired and the stored energy is exhausted by the total resistance of all the storage coils.

11.6 Projects and reality

A number of companies in the UK, USA, Germany, France, Japan and Russia started SMES R&D work in the early 1970 s. Since that time, many SMES projects have been proposed, but only some have been put into practice. The leading roles belong to the USA, Russia and Japan. As reported by the Soviet Academy of Sciences, the first Russian experimental SMES of 10^4 J energy capacity and with a rated power of 0.3 MW was connected, through a 6-pulse thyristor inverter, to the Moscow power system in the 1970s. This experimental SMES was constructed by the High Temperature Institute (IVTAN), which has subsequently been involved in a number of other SMES projects. Since 1989, this work has been done within the framework of the Russian State Scientific 'High Temperature Superconductivity' Programme. IVTAN's latest achievement is a 100 MJ 30 MW SMES installed in the Institute's experimental field, and is connected to the nearby 11/35 kV substation owned by Moscow Power Company. The electrical proximity of the 22 MW and 100 MW synchronous generators, as well as a specially designed load simulator, provides possibilities to conduct full-scale experiments on an SMES's influence on power system behaviour under normal and fault conditions.

The main particulars of this SMES design are given in Table 11.3. The superconducting coil is able to either store or release 50 MJ in normal mode, and up to 100 MJ in a so-called forced regime; its time constant is up to 3 s. In order to extend the experimental facilities, the coil design permits variations of the SMES

Table 11.3 Leading parameters of small-scale SMES projects

Parameter	LASL project	IVTAN project	
		Rated	**Forced**
Energy capacity (MJ)	30	50	80–100
Rated power (MW)	10	20	32
Maximal current in the winding (kA)	5.0	3	7.1
Available energy under periodic oscillations at 0.35 Hz	11	18	25–30
Bridge number	2	4	4
Bridge voltage (kV)	2.5	1	1
Bridge current (kA)	5	5	8
Working temperature (K)	4.2	4.2	4.2
Inductance (H)	2.4	11.1	3.96
Magnetic field (T)	3.92		
Mechanical pressure (MPa)	280		
Mean radius (m)	1.29		
Height (m)	0.86		
Current density (A/m^2)	1.8×10^9		
Refrigerator load (W)	150		
Liquid helium consumption (m^3/h)	$1{\cdot}5 \times 10^{-2}$		

parameters over a wide range. For that purpose, the coil is sectioned and each section can be connected with the others in either parallel or series.

The line-commutated, thyristor-based reversible convertor comprises four 6-pulse Greatz bridges, each operating at 1 kV 5 kA, which permits overloading up to 8 kA. Each Greatz bridge is connected to the grid through its own AC transformer. This scheme provides the required flexibility and consumes minimal reactive power.

If the attractiveness of such devices is confirmed, a commercial pilot SMES construction, the main details of which are given in Table 11.4, will be proposed for different roles in a power system.

Table 11.4 SMES for load levelling

Parameter	LASL project	IVTAN project
Energy capacity (GJ)	46 000	3600
Maximal current (kA)	50	110
Rated power (MW)	2500	500
Discharge duration (h)	5.1	2
Nominal voltage (kV)	110	220
Inductance (H)	37 000	595
Magnetic field (T)	4.6	3.5
Working temperature (K)	1.85	4.2
Coil diameter (m)	300	890
Coil height (m)	100	8.8
Coil thickness (m)	1.3	1.59

The first SMES for both experimental and commercial use has been designed by LASL and built for the Bonnevile Power Company in 1982. It was in use for some five years and was then dismantled for research purposes. Its main parameters are given in Table 11.3 and may readily be compared with the IVTAN scheme. The 30 MJ unit has been used as a stabiliser for power system damping of oscillations in a 1500 km long transmission line.

LASL has also proposed a large SMES for load-levelling purposes, as shown schematically in Figure 11.10. Its parameters are given in Table 11.4 and can be compared with the Russian pilot project.

According to the LASL reports, the costs of SMES construction are roughly distributed as follows:

- superconducting winding, 45%;
- support construction, 30%;
- on-site assembly, 12%;
- convertor, 8%; and
- cooling system, 5%.

ACCEL Instruments GmbH team has designed a 2 MJ SMES. The SMES is being used to ensure the power quality of a laboratory plant of the Dortmunder Elektrizitäts und Wasserwerke. The SMES is connected by a DC link convertor to the electrical grid of the plant. The system is designed to work for 8 s in a charge/discharge regime at 200 kW average power. The SMES CS coil is made of NbTi mixed matrix superconductor with helium transparent winding to improve the removal of AC losses of the superconductor during ramping of the magnet. Superconducting HTc current leads are applied to reduce the heat load from 300 K to the LHe level. This magnet system is designed for zero boil off of the LHe. The helium will be recondensed with a two-stage cryocooler. SMES has a sophisticated quench protection system.

Since the major portion of the expense is due to the cost of the superconducting winding, it appears extremely attractive to apply the effect of high-temperature (HT) superconductivity to SMES design. Unfortunately, the current-carrying ability of these HT materials is still too small for use in SMES construction. In addition, recent publications by Musuda in Japan and Bashkirov in Russia have shown that, although HT superconductors may introduce a reduction in thermal and insulating problems, the economic impact will be small, since capital cost savings are only 3–8% of the total capital cost.

Chapter 12
Energy storage in the power system itself

12.1 Power system as a flywheel

It should be mentioned here that all these devices are artificial secondary storage. Surprisingly, however, the power system itself, if properly controlled, can also act as a storage device itself, without any additional investment.

If there is any change in electricity demand in the power system, it is first accompanied by a slight drop in voltage as energy is extracted from, or supplied to, the electricity grid's equivalent capacitance while equivalent inductance is trying to maintain the current unchanged. The power system acts as capacitor bank storage, but with limited opportunities, since there are certain requirements on voltage deviation. The same concept may be applied to the power system's ability to act as a magnetic energy storage device.

The amount of energy stored in the grid's electromagnetic field is substantial but can only be used within tens of milliseconds. After that time, if the change in electricity demand continues, the frequency starts to deviate. This means that energy is extracted from, or supplied to, the rotating parts of the generating system – the power system acts as a flywheel storage device. This property of the power system, however, is limited by the permissible frequency deviation, so one can use only a small part of the energy accumulated in the rotating machinery. However, because there are many generators, the capacity of such a flywheel is substantial enough to cover changes in load demand for a few seconds.

The volume of stored energy available in Russian, Ukrainian and Kazakhstan utilities due to the permitted frequency deviation is shown in Table 12.1.

The main feature of these kinds of storage, being series-connected to the system, is that they react immediately to any change in load demand, providing an adequate power response.

If the frequency deviation exceeds that permitted by the power system's regulations, the steam governor opens or closes the valves and additional energy is extracted from, or supplied to, the enthalpy of the steam in the thermal power station's boilers. There is enough thermal energy stored in the boilers to cover changes in load demand for a few minutes, so the power system, too, is able to act as a thermal store.

To summarise the above, a power system has an ability to act as a capacitor, magnetic, flywheel or thermal energy storage device without additional investment;

Table 12.1 Power system as a flywheel

Utility	**Installed capacity**	**Energy discharged due to frequency decrease from 50 Hz to**		
		49.8 Hz f = **0.2 Hz***	**49.6 Hz** f = **0.4 Hz†**	**49.0 Hz** f = **1 Hz‡**
	GW	GJ	GJ	GJ
Central	53.4	1.31	2.61	6.50
Mid Volga	21.6	0.69	1.37	3.41
Urals	40.4	0.95	1.89	4.69
North-West	32.8	0.87	1.74	4.31
South	55.5	1.34	2.67	6.63
North Caucasus	10.7	0.31	0.62	1.55
Kazakhstan	12.0	0.28	0.56	1.40
Siberia	42.4	1.34	2.68	6.67
United system	268.8	7.52	1.50	3.73

According to the electricity quality regulations:
*Normal deviation.
†Maximal allowed deviation.
‡Deviation allowed for 90 h per annum.

generators play the role of power transformation systems, while thermal equipment, rotating machinery and transmission lines play the role of a central store. The capacities of these stores are limited, however, and therefore the power system's built-in storage can only accommodate short time fluctuations in load demand. However, this fundamental power system property allows small quantities of intermittent renewable sources to be accommodated on an interconnected grid without any technical problems.

12.2 Interconnected supergrid

The situation may be changed considerably if the so-called longitude effect is used. As is well known, there is an hour's time difference for each 1 000 km of longitude. The shape of the daily load curve suggests that, for load-levelling purposes, the beginning of an energy storage charge regime must be from 2 to 8 h away from the required beginning of the discharge regime. If there is a power system comprising, say, two regions, one of which has predominantly nuclear or thermal-based generation capacity while the other has predominantly hydro-based generation, and if these regions, being at least 2 000 km away from each other, are connected by a powerful enough transmission line, it is possible that the system can have a built-in pumped hydro storage that can be used for load levelling within the interconnected system.

To illustrate this, let us consider an interconnected system comprising two separate parts where, owing to the time difference, peak demands do not occur simultaneously. When peak demand in hydro-based system 1 starts (or is already in process, depending on the time difference between the two systems), it is already a demand trough in the thermal or nuclear-based system 2. In that case, water required for energy generation can be stored in the hydro plant's reservoirs in system 1, while the necessary energy will be generated by spare capacity of thermal or nuclear plants in system 2 and transmitted to system 1. This is a charge mode of the built-in pumped hydro storage system.

When peak demand in system 2 starts the thermal and nuclear plants, being constantly loaded, are feeding consumers in system 2 while hydro plants in system 1 (where the demand trough occurs), using the stored water, are supplying peak energy to system 2. This is a discharge mode of the built-in pumped hydro storage.

In such a built-in pumped hydro system, the transmission lines act as waterways in a real pumped hydro system, the thermal or nuclear power plants play the role of pumps, and the hydro plant itself acts as a generator while its reservoir is used as a central store.

European and Russian utilities, were they interconnected by powerful transmission lines, would provide the possibility of exploiting built-in pumped hydro properties. The majority of generating capacity in the European utilities (Russia included) is concentrated on base-load coal-fired or nuclear power plants, whereas in eastern Siberia and northern Europe more than 80% of the installed capacity is concentrated in large hydro plants. The time difference between London and Krasnoyarsk is 9 h. All utilities are electrically connected: a cross-Channel link connects the UK and France, which, via Germany and the so-called 'Mir' system, is connected to the Russian 'Central' utility and then is connected via the Urals to the 'Siberia' utility. This transcontinental connection link has never been used for this purpose, but theoretically it is possible to imagine the following scenario: when there is an evening peak demand in the UK, it is night-time trough demand in Siberia, so its hydro plants generate peak energy for British consumers. When there is peak demand in Siberia, there is a trough in demand in Europe. European base-load plants, instead of decreasing their load, supply energy to Siberia, where local hydro plants are storing water and generating only the environmentally constrained minimum to maintain river flow. The same may probably be applied to the northern Europe hydro plants and Russian 'Central' utility, since there is also a sufficient time difference and the relevant transmission lines could be erected.

The Siberian utility may also be connected to the North American power system. As is clear from Figure 12.1, the difference in the load curves owing to the significant time difference provides the possibility of using Siberian hydro plants for peak generation in the USA and using American base-load plants for peak load coverage in Siberia. The difference in the nominal frequencies could easily be overcome by using a DC transmission line.

The main technical problem to be solved is the necessity to transmit large amounts of power over long distances. The most promising type of line for this purpose, high-voltage direct current (HVDC) transmission lines, is under development in the USA, Canada and Russia.

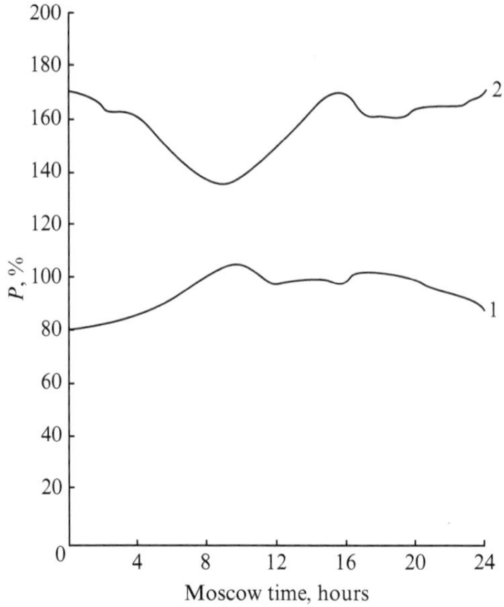

Figure 12.1 Comparison of load curves for (1) Russian and (2) US utilities

Chapter 13
Considerations on the choice of a storage system

13.1 Comparison of storage techniques

In summarising the information given in the preceding chapters, it is necessary to review the status of large-scale electrical energy storage and to make a comparison of storage plants with very different characteristics, as well as considering conventional alternatives.

Pumped hydro is the only type of storage with a well-developed and highly reliable technology. There is more than 50 GW in widespread commercial use in power systems around the world and no less than 10 GW is under construction. The main problem with this type of storage is that it is not always easy to find sites suitable for two reservoirs separated by at least 100 m, which are not remote from the power grid and yet have suitable physical characteristics. Massive civil engineering works are required, and since these locations are often in areas of scenic importance, great care has to be taken over the environmental effects of such schemes. For this reason, there are a number of proposals to establish one of the reservoirs hundreds of metres underground.

A conceptually simple way to store energy in a form convenient for power generation is to pump compressed air into an underground reservoir. Compared with pumped hydro this method has apparent advantages: the air storage cavern can be in either hard rock or salt, providing a wider choice of geological formation, and the density of the energy stored is much higher, that is, a smaller size for an economically viable installation. For a given volume of the underground reservoir, it would be better to store compressed air, since, to yield the same power output as a CAES, the reservoir for pumped hydro would have to be very much deeper and could encounter appreciable geothermal temperatures. When the cavern is constructed from salt, it is likely that the cost of extending the storage period of weekly storage will not be so great as with pumped hydro and other schemes.

It is clear that a CAES of comparatively small size (up to a million kWh) and short construction time (up to 5 years) presents a much smaller financial risk to a utility than even the smallest economically reasonable pumped hydro plant of approximately 10 million kWh. There is, however, one complication: since air gets hotter when it is compressed, it must be cooled before it is stored in order to prevent fracture of the rock or creep of the salt. The stored air must then be reheated by burning a certain amount of fuel as the air is expanded into the turbine that drives the electricity generator; the need for thermal storage is clear. Against CAES, there is also the need to use premium fuels like oil distillates or natural gas to power the gas turbine. It is possible to overcome this drawback using synthetic fuels – methanol, ethanol, ammonia or hydrogen – instead of natural ones. Methanol has half the volumetric energy density of petrol, is very corrosive and has a high temperature of vaporisation. Ethanol has cold start problems and therefore requires manifold heating. Methanol and ethanol are best used as petrol extenders, but it has been considered that methanol itself would make an excellent turbine fuel. Ammonia is the least suitable synthetic fuel for internal combustion engines.

Hydrogen, generated from water by electrolysis, could be stored as compressed gas, liquid or metal hydride, and reconverted into electricity either in a fuel cell or through the conventional gas turbine-generator chain. Hydrogen energy storage systems would have considerable flexibility with respect to the location and operation of the storage plant; its transportation exploits well-established technology for natural gas, and moreover, pipeline transmission over very long distances is cheaper and less objectionable on environmental grounds than electricity distribution.

The principal disadvantages of gaseous hydrogen as a storage medium are that it requires large volumes, it is explosive and it is difficult to ensure leak-tight vessels.

The inconvenience of a highly cryogenic and inflammable medium such as liquid hydrogen, the sophisticated engineering required for production and transfer to and from the storage vessels and the cost of these operations make liquid hydrogen-based storage rather unattractive.

There are some drawbacks to metal hydride hydrogen; the greatest being its weight and the high price of 'host metals'. For stationary energy storage, the price of the metal is the crucial factor, whereas, in the case of transportation and use as a fuel for vehicles, it is a combination of price and weight that forms the deciding factor. Overall hydrogen-based storage appears to be rather expensive, complex and relatively inefficient.

Today, the bulk of hydrogen is produced from low-cost oil and natural gas, and is used almost exclusively for chemical purposes such as the synthesis of ammonia, methanol, petrochemicals, and for hydrocracking within oil refineries. Hydrogen as fuel amounts to less than 1% of its annual production, and therefore, it is difficult at present to define the cost of hydrogen for large-scale applications. It is clear that it cannot at present compete economically with fossil fuels, and this situation is likely to continue until, owing to the current tendency of organic fuel prices to rise, alternative primary energy sources become substantially cheaper than fossil fuels. There are major technical problems to be solved in the production, utilisation and storage of hydrogen, but, nevertheless, it is the most promising concept for future environmentally benign power systems.

Thermal energy storage is either part of the thermal subsystem of a power system or a secondary source of heat for a consumer. In its first function, TES is not a 'standalone' device but has to be directly linked to the steam generator. Swings of up to 50% may be achievable with thermal energy storage designed to raise steam, which is passed through a peaking turbine to avoid overloading the main turbine plant. Long charge times are then desirable to ensure that the main turbine runs as close as possible to its design load. TESs are therefore ideally suited for load-levelling purposes in a power system. Coupling TES to a modern coal-fired plant requires the steam reheater to be eliminated from the cycle to avoid excess reheater tube temperatures. This implies a considerable power-related cost and performance penalty, which could be avoided either by designing coal-fired plant together with TES or using PWR plant for this form of storage.

The use of a peaking turbine during discharge periods, for the steam that normally performs feedheating duties, and the use of direct live steam extraction for charging the store minimise off-design steam flow variation in the main turbines. This requires TES installation in the basic cycle of feedwater heating, which will have limited flexibility since it is restricted to operation with power swings of less than 20%.

There are advantages and disadvantages of being an integral part of a thermal power plant. Apart from the requirement to be less expensive than peaking capacities, the following practical issues have to be carefully evaluated: plant safety, availability, reliability, flexibility, stability of operation and maintenance.

Various TES concepts have been studied in conjunction with the thermal power station steam cycle, and some have been used for many years, but according to modern calculations for steam generation, only pressurised water in lined underground caverns and above-ground oil/rock heat storage in atmosphere pressure vessels are found to be economically viable.

Thermal energy storage is an essential part of a modern CAES concept, and is well known to many of the population of the Northern Hemisphere as a source of secondary heat at a consumer's premises. It also becomes extremely viable as a store of cold, and in this capacity is very competitive.

The other types of storage equipment, such as flywheels, chemical batteries, capacitor banks and SMES, give the following common advantages:

- They are environmentally benign with no requirements for cooling water, no air pollution, minimal noise and moderate siting requirements.
- They have extremely small power reverse times so that power can be delivered or consumed practically on demand, offering increased flexibility in meeting area requirements.
- Power reverse capability can aid in responding to emergency conditions.

Flywheels are under active development, mainly for vehicle applications and for impulse power generation for large-scale storage applications; it seems likely that they will be provided in relatively small modules. They offer a number of attractions for energy storage. First, for short durations of the charge-store-discharge cycle, they are highly efficient. Second, due to limitations of their materials, they are only available in relatively small modules, but this drawback for large power systems is an advantage for small-scale applications – they can be made at different sizes and can be installed anywhere they are required in an electricity distribution system. They do not constrain the number and frequency of charge–discharge cycles, and they are environmentally benign. Flywheels are capable of absorbing and releasing energy quickly but recent studies indicate that, even using advanced designs, they will remain too expensive for large-scale power system applications. Within power system applications, the field of usage for flywheels is likely to be in distribution.

Chemical batteries are very attractive for a number of storage applications in both the supply and demand sides of a power system. Batteries have the following excellent properties:

- They store and release electrical energy directly.
- Being modular, they can be used flexibly.
- They are largely free of environmental problems.
- They have no mechanical ancillaries.
- They can typically have a short lead time in manufacture.

Estimates show that currently available $Pb\text{-}PbO_2$ and NiO-Fe batteries have costs commensurate with target figures for load levelling with a nuclear base-load. Many other systems, such as Na-S and $Zn\text{-}Cl_2$ batteries, which are both at an advanced stage of development, $Zn\text{-}Br_2$, Li-FeS and the solid-state $Li\text{-}Sb_2S_3$, show possibilities for considerable savings, and are therefore promising. Also promising is solid Li_4Si coupled with $TiS_2\text{-}Sb_2S_3\text{-}Bi$, the solid electrolyte being LiI dispersed in fine Al_2O_3 powder. Containment of reactive materials would be accomplished in the battery itself and corrosion should be eliminated. Thin-film technology used in voltaic pile construction will provide the basis for production of this low-cost system.

The modular construction of batteries permits factory assembly, which leads to low site costs and short construction lead times. Since chemical batteries have a minimal environmental impact associated with their modest site requirements, high safety, low pollution and low noise, this allows them to be dispersed optimally in essentially small units close to the consumer, thereby providing savings in transmission costs compared with other schemes.

Batteries situated close to the consumer are able to smooth the load on the distribution network, thus decreasing the required capacity of substations. They could also be used as an additional source of thermal energy for a district heating scheme, utilising waste battery heat generated during the daily charge–discharge cycle, thus working together with TES at the consumer's premises. Considerable research and development effort is aimed towards creation of a low-cost, high-energy density and reliable electric vehicle battery.

Battery storage for solar electricity is one of the fastest developing areas of application. In most present applications, a rechargeable chemical battery is linked with solar cells to cover periods of insufficient sunshine. Reduction of production costs for mass-produced solar cells have raised significant interest in solar-cell batteries.

Like batteries, capacitor banks have a modular structure and therefore allow factory assembly of standard units, providing a short lead time from planning to installation and thereby reducing capital costs. The main drawback of capacitor bank storage is its low energy density in comparison with batteries, but since capacitors have very low internal resistance, the power density is very high; they could be used for power multiplication where necessary.

Superconducting magnetic energy storage schemes can store electricity directly, and therefore with high efficiency, but they are still extremely expensive. SMES only seems to have the potential to become economically attractive on a very large scale. Although this is a high-technology area, there are no insoluble

Table 13.1 Comparison of storage alternatives

Type	Efficiency (%)	Relative power component	Capital cost energy component	Reasonable energy capacity (J)	Energy density (J/m^3)	Construction lead time (years)	Lifetime (years)	Number of cycles	Reverse time (s)	Siting suggestion
Flywheel	85	0.7	30	10^9	10^8	3	20	Unlimited*	0.1	Close to consumer's load terminal
Pumped hydro†	80	1.0	1.0	10^{13}	10^{6}‡	8	50	Unlimited*	10	Geological considerations
CAES	f_{cef}* = 1.3 f_{fhr} = 4300 kJ/kWh	2.1	0.4	10^{12}	10^5	3	25	Unlimited*	360	Geological considerations
Hydrogen as a synthetic fuel†	50	8.6	0.6	10^{12}	10^9	3	25	Unlimited*	360	Close to a gas turbine plant
TES	75	2.5	3.0	10^{11}	10^9	12	30	Unlimited*	Tens of minutes	Part of the thermal plant
Batteries	80	0.6	6.0	Not constrained	10^8	2	10	500	0.01	Close to consumer's load terminal
Capacitor	80	0.6	5.5	Not constrained		2	10	10^7	0.01	Close to consumer's load terminal
SMES	90	0.6	14.0	10^{13}		12	30	10^6	0.01	Power system's substations, generator's terminal

*Not limited within expected lifetime.
†Electrolysis + gas turbine.
‡For a 100 m water head.

technical problems. However, the necessity to work with large units requires experience, which can only be gained by exploiting small and economically unjustifiable devices. The development and launch costs will therefore be high.

The main quantitative details of different storage equipment are given in Table 13.1. One may conclude that such forms of storage as TES, CAES and pumped hydro have a relatively large reverse time, and therefore their possible applications are limited compared with flywheels, chemical batteries and SMES whose reverse time is very small. On the other hand, as is clear from Figure 13.1, TES, CAES and pumped hydro are suitable for large-scale applications in power systems while flywheels, owing to their limited size, or chemical batteries and capacitor banks, owing to their modular construction, are better for comparatively small-scale applications at the supply side, or as dispersed storage at the demand side, of the power system. Only SMES can be used anywhere in the power system but economic considerations make the prospects for application of this technically very attractive type of storage rather remote. Figure 13.2 shows possible locations for energy storage installations in the power system.

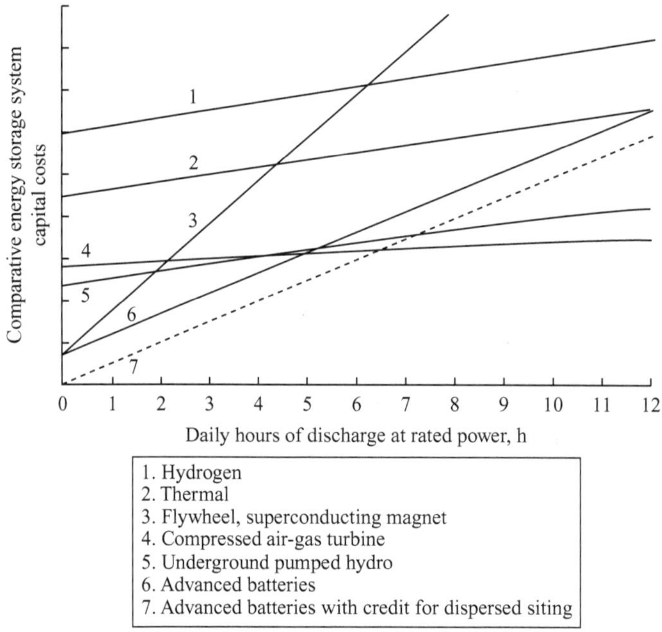

Figure 13.1 Costs of energy storage systems

From the environmental point of view the CAES concept looks very attractive. It includes hydrogen generated from renewable energy sources, which is then used as a storage medium and for fuel for the CAES. Unfortunately it is somewhat expensive at present, but fuel price rises could improve this position. More promising still is the idea of combining different types of storage

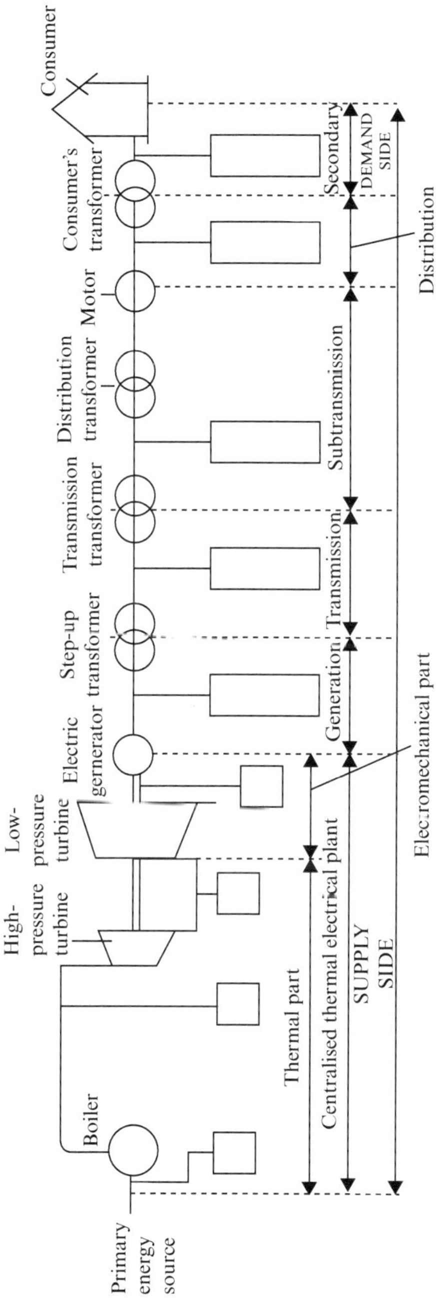

Figure 13.2 Typical electricity utility power system with dispersed storage

equipment to make better use of their properties. The first example should be an adiabatic CAES of which TES is an essential part. A combination of CAES and flywheel or pumped hydro and chemical batteries, in order to get large-scale storage, on the one hand, and quick response, on the other, is also promising. However, all of them have to be justified economically: the cheapest and simplest of all could be the storage of additional hot water in the boiler circuit of a thermal power plant.

Energy storage usage is being restrained by a lack of regulation to guide utilities and investors. There is no overall strategy of the incorporation of energy storage technique and methods into the power systems particularly in those with separation of transmission and generation. Energy storage benefits both generation and transmission companies, and this immediately leads to confusion concerning the question of energy storage control. This situation leaves utilities and investors uncertain as to how investments in energy storage should be treated and how costs could be recovered, so neither is willing to invest. The technology should be given more financial and institutional support – alike renewable energy technologies – in order to encourage the construction of storage plant.

A slightly simplified rule states that storing heat saves energy while storing electricity saves capital investment. Since the energy industry on a global basis faces a resource crisis, both in respect of primary energy and capital investment, there are good reasons for investigating the storage of low-quality as well as high-quality energy.

Part III

Power system considerations for energy storage

Chapter 14
Integration of energy storage systems

14.1 Problem formulation

Energy storage is necessary because the demand side in a power utility is characterised by hourly, daily and seasonal variations, whereas the installed capacity of the supply side is fixed. To facilitate this varying demand at minimum cost and acceptable reliability, the utilities plan and operate their generation resources to match the load characteristics. During the decision-making process of planning, information regarding the effect of an energy storage unit on power system reliability and economics is required before it can be introduced as a decision variable in the power system model. The main objectives of introducing energy storage to a power utility are to improve the system load factor, achieve peak shaving, provide system reserve and effectively minimise the overall cost of energy production. Constraints of various systems must also be satisfied for both charge and discharge storage regimes.

Factors related to the cost of this newly introduced unit have to be considered, including an economic evaluation of its performance characteristics. The impact of dispersed energy storage integrated within the system has to be considered, including the effects of distributed units on system stability and spinning reserve requirements.

The economics of storage devices have an influence on both the initial capital investment in the system and the operating and maintenance costs. This is shown schematically in Figure 14.1, from which it is clear that there is a certain storage rated power (and also energy capacity) usage, which leads to minimal cost of electricity. To provide a low-cost electricity service, the life-cycle cost of an energy storage system must be competitive with more conventional power sources, for peaking and intermediate applications, such as gas turbines and combined cycle units. Turnaround efficiency and the expected lifetime also have an effect on the economic evaluation. Successful integration of energy storage into the utility grid depends on ES type and design, balance of system requirements and associated capital and operating costs.

The effects of introducing energy storage into a power system originally designed without it are not merely financial; all aspects of generation expansion planning need to be considered. A key aspect of this work is evaluation of cost benefits for varying levels of energy storage system penetration and sensitivity analysis of the following parameters:

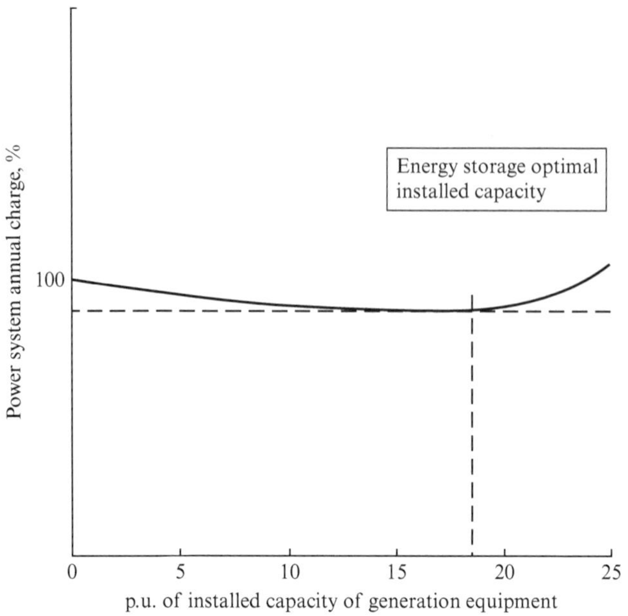

Figure 14.1 Power system annual charge with respect to installed capacity of the energy storage

- system average daily load factor and load density factor;
- capital costs for power equipment, transmission lines and storage means and escalation;
- fuel price and its escalation; and
- energy storage efficiency.

The addition of energy storage to the power system will certainly lead to a new mix of generation structure. ES will possibly displace part of the peaking capacity and will require more base capacity, which in turn will displace intermediate capacity. Any composition of the generating units in the supply side of the power system has to satisfy the main requirement of the demand side: to cover its load curve.

Let us consider that, for coverage of the load curve, there are installed: N_{b0} = capacity of base power sources, N_{i0} = capacity of intermediate power sources and N_{p0} = capacity of peak power sources. Storage is not included in this structure at all. Let us increase the base capacity by δN_b. This capacity will generate excess energy E, which may be calculated as

$$E_c = \sum_{i=1}^{t_h} (N_{b0} + \delta N_b - L_i) t_i$$

where t_h = duration of demand trough.

This excess energy could be used for storage charge and then for discharge during peak demand, thus decreasing the capacities required for peak and intermediate power sources. The installed capacity of the intermediate power sources may be decreased by the same δN_b.

The peak capacity could be decreased by energy storage discharge capacity P_{sd}, which can be calculated on the basis of energy balance in the storage unit:

$$P_{sd} = \frac{E_c}{(\xi_s t_p)}$$

where t_p = duration of peak demand.

So the introduction of storage leads to a completely new mix of the supply side structure; this change of structure can be given by the following set of equations:

$$N_{b1} = N_{b0} + \delta N_b$$
$$N_{I1} = N_{I0} - \delta N_b$$
$$N_{p1} = \frac{N_{p0} - \delta N_b t_h}{(\xi_s t_p)}$$

If the storage is sited close to the power plants then the above structural changes are the only ones necessary.

Since for certain storage techniques there is the possibility to site them close to the consumers, all the power flow in the transmission lines will be changed.

If, for example, the energy storage is placed at the end of the ith transmission line, the introduction of storage will also affect substation and transmission loads. These changes will lead to changes in energy losses (hopefully decreases).

It is clear from the above that the introduction of energy storage into the power system may completely change the structure of its supply side.

So we have a clear optimisation problem that can be formulated as follows: minimise the annual charge for the power system as a function of the installed capacities of its elements and its fuel consumption, subject to the set of system constraints that may be written in the form:

$$\min f(x) = C\mathbf{X}$$
$$\text{subject to } \mathbf{A}x < b$$

where $f(x)$ = power system cost function
x = vector of variables
A = matrix, resulting from the set of constraints
b = right-hand sides of constraints equations.

It is therefore clear that, for quantitative evaluation, we need a special mathematical model of the power system including energy storage.

14.2 Power system cost function

To start, the model has to be represented by a nonlinear annual charge dependence on power system capital cost, amortisation cost, price for the fuel necessary to generate a predetermined quantity of energy, including losses in the storage, and above all a storage capital cost.

Each of the terms mentioned is a function of a number of variables; let us consider them carefully.

The capital cost of a power system K_{pws} comprises the cost of power plant K_{pp}, cost of substations K_{ss} and cost of transmission lines K_{tl}:

$$K_{pws} = K_{pp} + K_{ss} + K_{tl}$$

The capital cost of power plants on the supply side of the power system depends on the generation structure and may be given by

$$K_{pp} = \sum_{i=1}^{n_1} K_{bi}^{*} N_{bi} + \sum_{i=1}^{n_2} K_{Ii}^{*} N_{Ii} + \sum_{i=1}^{n_3} K_{pi}^{*} N_{pi}$$

where $K_{bi}^{*}, K_{Ii}^{*}, K_{pi}^{*}$ = specific cost per unit of rated capacity for ith base, inter intermediate and peaking power sources, respectively

N_{bi}, N_{Ii}, N_{pi} = standard installed capacity for ith power source used for base, intermediate and peak load coverage.

The necessary reserve capacity N has to be included in the power system installed capacity; therefore

$$N_{pws} = \sum_{i=1}^{n_1+n_2+n_3} N_i + N_r = L_b + N_{rb} + L_i + N_{ri} + L_p + N_{rp}$$

where L_b, L_i, L_p = base, intermediate and peak parts of the load curve, respectively

N_{rb}, N_{ri}, N_{rp} = reserve capacity installed on the base, intermediate and peaking equipment, respectively.

For long term planning purposes, reserve capacity evaluation is usually simplified and defined as 13–20% of the power capacity required for load coverage. It should be mentioned that the reserve capacity has to be more than the installed capacity of the largest unit in the reference utility system.

The capital cost of substations depends on several factors: substation voltage class, rated power flow, substation type and many others. Introduction of energy storage affects power flow in the system and thereby affects the capital cost of substations. Therefore there are variable and constant parts of the substation capital cost:

$$K_{ss} = \sum_{i=1}^{p} C_{ssi} + \sum_{i=1}^{p} K_{ssi}^{*} P_{i\max}$$

where p = number of substations in the power system

C_{ssi} = constant part of the ith substation (voltage and scheme)

K_{ssi}^{*} = specific cost per unit of rated capacity of the ith substation

$P_{i\max}$ = rated capacity (maximal calculated power flow) of the ith substation.

The capital cost of the transmission lines is a significant part of the overall cost and therefore must be included. But in the first stage of the planning process, it is possible to simplify the corresponding part of the power system model using specific cost K^*_{tl} per kilometre of length and per megawatt of rated power flow.

The transmission line model may then be given as follows:

$$K^*_{tl} = \sum_{i=1}^{m} K^*_{\mathrm{tl}i} l_i P_{\mathrm{tl}i}$$

where $l_i, P_{\mathrm{tl}i}, K^*_{\mathrm{tl}i}$ = length, rated power flow and specific capital cost per kilometre per MW of the ith transmission line, respectively.

All the above-mentioned elements of the power system econometric model are independent of the electrical regime of the reference power system and comprise the constant part of the annual charge. The variable part comprises fuel cost, cost of compensation for energy losses (usually included in the fuel cost), amortisation and maintenance costs, staff salaries, etc.

Maintenance cost also depends on the loading of thermal units, but this dependence is still somewhat uncertain, so in the present model this dependence has not been taken into account. Therefore, maintenance and amortisation costs, together with salaries, are represented in the model as a linear capital cost dependence and are given as follows:

$$A = \sum_{i=1}^{n_1} a_{\mathrm{b}i} K^*_{\mathrm{b}i} N_{\mathrm{b}i} + \sum_{i=1}^{n_2} a_{1i} K^*_{\mathrm{l}i} N_{\mathrm{l}i} + \sum_{i=1}^{n_3} a_{\mathrm{p}i} K^*_{\mathrm{p}i} N_{\mathrm{p}i}$$
$$+ \sum_{i=1}^{p} a_{\mathrm{ss}i} C_{\mathrm{ss}i} + \sum_{i=1}^{p} a_{\mathrm{ss}i} K^*_{\mathrm{ss}i} P_{\mathrm{max}i} + \sum_{i=1}^{m} a_{\mathrm{tl}i} K^*_{\mathrm{tl}i} l_i P_{\mathrm{tl}i}$$

where $a_{\mathrm{b}i}$, $a_{\mathrm{l}i}$, $a_{\mathrm{p}i}$, $a_{\mathrm{tl}i}$, $a_{\mathrm{ss}i}$, = amortisation coefficients for the ith base, intermediate and peak unit, transmission line and substation, respectively.

Fuel cost is more than 60% of this variable part and depends on the regime of the power plant units, and particularly on the fuel consumption curves of each unit and their current load. The problem is that the fuel consumption curve is a linear function of the unit's load only to a first order. In reality, it has a nonlinear dependence, and to calculate the influence of energy storage on the reference power system economics it is necessary to take this nonlinearity into account.

Energy losses in the transmission line depend on their regimes too, and are nonlinear functions of the corresponding power flow of line current. In the present model the fuel consumption curve is represented by a quadratic approximation and may be given as follows:

$$b_{\mathrm{f}} = A(N - N_{\mathrm{min}})^2 + B(N - N_{\mathrm{min}}) + C$$

where b_{f} = fuel consumption, tons/h
N = current generated electric power, MW
N_{min} = technically limited minimal generated power
A, B, C = approximation coefficients, different for different types of the power units. Table 14.1 reflects this difference for the power units installed on the Russian power systems.

This approximation is only valid for base and intermediate coal-, oil- or gas-fired power plants. The fuel consumption for nuclear plants is assumed to be constant

Table 14.1 Approximation coefficients for the fuel consumption curves

Fuel type	Rated capacity (MW)	Minimal load (MW)	C	Coefficients	
			t	B (t/MW)	A (t/MW^2)
Oil	800	350	120.44	0.266	$0{\cdot}496 \times 10^{-4}$
Oil	300	150	41.16	0.239	0.65×10^{-4}
Coal	300	180	56	0.3	0.36×10^{-4}
Gas	200	120	42	0.295	0.5×10^{-4}
Oil	200	120	38	0.330	0.15×10^{-2}
Coal	200	120	48	0.259	0.85×10^{-3}

because of their constant loading. Fuel consumption for gas turbines is a linear function of their load. For hydro plants, and practically all kinds of renewable sources (except small biomass-fired thermal installations), fuel consumption and corresponding fuel cost is not applicable.

The cost of fuel, K_{f}, required for load curve coverage for a given period of time, may be given by the following expression:

$$K_{\mathrm{f}} = \sum_{j=1}^{n} \sum_{i=1}^{T} C_{\mathrm{f}j} b_{\mathrm{f}j}(N_i) t_i$$

where C_{f_j} = specific fuel cost per ton for the jth unit
b_{f_j} = fuel consumption for the jth unit
N_i = current generated power
t_i = time duration when N_i is constant
T = period of time
n = number of units.

Obviously, the econometric model of energy storage itself also has to be included in the power system model. In the form of annual charge, the overall power system econometric model can be written as follows:

$$f(x) = R_{\mathrm{pp}} K_{\mathrm{pp}} + R_{\mathrm{ss}} K_{\mathrm{ss}} + R_{\mathrm{tl}} K_{\mathrm{tl}} + A + K_{\mathrm{f}} + R_{\mathrm{es}}(K_{\mathrm{p}}^{*} P_{\mathrm{s}} + K_{\mathrm{e}}^{*} E_{\mathrm{s}})$$

where each different R is the rate of return required on the corresponding capital cost K. In full form, using all the above-mentioned expressions:

$$\begin{aligned} f(x) = {} & R_{\mathrm{pp}} \left(\sum_{i=1}^{n_1} K_{\mathrm{b}i}^{*} N_{\mathrm{b}i} + \sum_{i=1}^{n_2} K_{\mathrm{l}i}^{*} N_{\mathrm{l}i} + \sum_{i=1}^{n_3} K_{\mathrm{p}i}^{*} N_{\mathrm{p}i} \right) \\ & + R_{\mathrm{ss}} \left(\sum_{i=1}^{p} C_{\mathrm{ss}i} + \sum_{i=1}^{p} K_{\mathrm{ss}i}^{*} P_{i\max} \right) + R_{\mathrm{tl}} \sum_{i=1}^{m} K_{\mathrm{tl}i}^{*} l_i P_{\mathrm{tl}i} \\ & + \sum_{i=1}^{n_1} a_{\mathrm{b}i} K_{\mathrm{b}i}^{*} N_{\mathrm{b}i} + \sum_{i=1}^{n_2} a_{\mathrm{l}i} K_{\mathrm{l}i}^{*} N_{\mathrm{l}i} + \sum_{i=1}^{n_3} a_{\mathrm{p}i} K_{\mathrm{p}i}^{*} N_{\mathrm{p}i} \\ & + \sum_{i=1}^{m} a_{\mathrm{tl}i} K_{\mathrm{tl}i}^{*} l_i P_{\mathrm{tl}i} + \sum_{i=1}^{p} a_{\mathrm{ss}i} (C_{\mathrm{ss}i} + K_{\mathrm{ss}i}^{*} P_{i\max}) \\ & + \sum_{j=1}^{n_1+n_2+n_3} \sum_{i=1}^{T} C_{\mathrm{f}j} b_{\mathrm{f}j}(N_i) t_i + (R_{\mathrm{es}} + a_{\mathrm{es}}) K_{\mathrm{e}}^{*} E_{\mathrm{s}} + (R_{\mathrm{pts}} + a_{\mathrm{pts}}) K_{\mathrm{p}}^{*} P_{\mathrm{s}} \end{aligned}$$

This econometric model of the supply side, including storage, may be used for power system analysis only if the set of system constraints is formulated in the form of equations and non-equations.

14.3 System constraints

All the power system regime parameters mentioned in the econometric model have to satisfy certain requirements of the demand side or system constraints.

First of all, the power balance has to be achieved in the total system and in each node. This means that, for any ith moment of time, the energy balance equation has to be satisfied:

$$\sum_{j=1}^{n_1+n_2+n_3} N_j \pm P_s - \sum_{i=1}^{m} \delta P_i - \sum_{l=1}^{k} L_i = 0$$

In the above formula, N_j represents the power generated by a corresponding unit that is more than the consumer requires. The difference between the generated and consumed power is the so-called power losses:

$$\delta P_j = \sum_{i=1}^{n} N_i - \sum_{i=1}^{k} L_i$$

where n, k = number of generation units and consumers, respectively
δP_j = Power losses during the jth period of the reference time.

Since information on the demand side of the power system load curve is the main data available, it is necessary to calculate power losses; otherwise, it is impossible to obtain the required power generation. Power losses in the average utility grid are approximately 7–12% of the total load demand, and they depend mostly on the electrical regime of the power system.

The location of energy storage has a considerable influence on the power flow in the utility grid, so power losses depend on the site and regime of the ES, and must therefore take into account the siting of the storage concerned.

Power losses in a utility grid are the sum of corresponding losses in all its elements. The two different types of grid elements usually considered are transmission lines and transformers. Power losses in the transmission lines, δP_{tl}, may be given by

$$\delta P_{\text{tl}} = \delta P_{\text{cor}} + \delta P_{\text{wr}} = U^2 g_{\text{cor}} + \left(\frac{S^2}{U^2}\right) r_{\text{tl}}$$

where δP_{cor} = losses due to corona phenomena
δP_{wr} = losses in the wires of the transmission line
U = network voltage
g_{cor} = effective conductance associated with corona losses
r_{tl} = resistance of the line
S = total power (active and reactive) at the demand end of the line.

Power losses in the transformer, $\delta P_{\rm tr}$ may be given by

$$\delta P_{\rm tr} = \delta P_{\rm oc} + \delta P_{\rm w} = \frac{\delta P_{\rm oc} + \delta P_{\rm sc} S^2}{S_{\rm rt}^2}$$

where $\delta P_{\rm oc}$, $\delta P_{\rm sc}$, $\delta P_{\rm w}$ = open circuit, short circuit and wiring power losses, respectively, in the transformer
S, $S_{\rm rt}$ = current and rated apparent power via transformer.

For ease of computation, both these expressions are usually simplified: linearisation is required to use standard LP packages.

The installed capacity of all the generating units, including the energy storage, has to be larger than the maximal load demand:

$$\sum_{i=1}^{n_1} N_{{\rm b}i} + \sum_{i=1}^{n_2} N_{{\rm l}i} + \sum_{i=1}^{n_3} N_{{\rm p}i} + P_{\rm s} > \sum_{i=1}^{k} L_i$$

Since the energy storage is charged during the night-time trough, the installed capacity of the base units minus storage capacity has to be larger than the minimal load demand:

$$\sum_{i=1}^{n_1} N_{{\rm b}i} - P_{\rm sc} > \sum_{i=1}^{k} L_{i{\rm min}}$$

The installed capacity of all the base units has to be less than the average load demand:

$$\sum_{i=1}^{n_1} N_{{\rm b}i} < \frac{1}{T} \sum_{i=1}^{T} L_i t_i$$

The isoparametric conditions for hydro plant water consumption $\theta_{\rm kg}$, and fuel consumption for certain thermal power plants, $B_{\rm mg}$, are

$$\sum_{j=1}^{m} \theta_{{\rm k}j}(N_{{\rm k}j}) - \theta_{\rm kg} = 0$$

$$\sum_{j=1}^{m} B_{{\rm m}j}(N_{{\rm m}j}) - B_{\rm mg} = 0$$

The energy balance in the storage for a given period of time T is

$$\sum_{j=1}^{m} (E_{{\rm c}j} - E_{{\rm d}j} - \delta E) = 0$$

The energy storage power limitations for each time interval are

$$P_{c\,\min} \leq P_c \leq P_{c\,\max}$$
$$P_{d\,\min} \leq P_d \leq P_{d\,\max}$$

The storable energy limitations of energy storage for the *j*th charge–discharge cycle are

$$E_0 \leq E_{dj} \leq E_s$$
$$0 \leq \sum_j E_{cj} \leq E_s$$

The *i*th power plant capacity limitations for each time interval are

$$N_{i\min} \leq N_i \leq N_{i\max}$$

where $N_{i\min}$, $N_{i\max}$ are the minimal and maximal capacity technically permitted for the *i*th power plant.

The permitted voltage limits in the *i*th node for each time interval are

$$V_{i\min} \leq |V_i| \leq V_{i\max}$$

The heating conditions for the transmission line wires are

$$|I_t| \leq I_{l\,\text{perm}}$$

where $I_{l\,\text{perm}}$ = maximal permitted current through the *l*th transmission line.

The system stability limitations for the *l*th transmission line for each time interval are

$$|P_{tl}| \leq P_{tl\,\max}$$

where $P_{tl\,\max}$ = maximal allowed power flow through the *l*th line under power system stability conditions.

The economic model formulated above for a power system with storage, comprising the power system cost function and a set of system constraints, allows us to investigate possible structural changes in the supply side due to the introduction of energy storage and to find out the optimal mix of power sources and storage for a given power utility.

As is clear from the economic model, the optimal values of rated power and energy capacity of the storage unit to be introduced in the supply side structure depend on a number of parameters such as specific cost for base, intermediate and peak power equipment; fuel costs for that equipment; load curve parameters; and storage efficiency. In considering the economic value of storage, four main factors should be taken into account.

First, there is an enhancement factor related to the uniqueness of the energy made available by the storage unit. Second, there are generating capacity credits arising from the possible displacement of otherwise needed power sources by storage. Third, there are energy utilisation credits resulting from imported-power dispatch and energy management in consumption, including credits from postponement of system development. And, finally, there are production cost credits owing to use of lower cost fuel or higher generating efficiency being achieved by the use of storage.

These four factors represent the annual benefits of the use of storage, B_s, and in fact are the sum of the following:

- value of recovered energy;
- displaced capacity credit;
- utilisation and postponed investment credit;
- maintenance credit; and
- energy production cost credit.

The introduction of energy storage is justified if these annual benefits are no less than the annual charge for storage. Using the storage economic model, we can obtain the target specific cost per rated power as a function of its target specific cost per energy capacity:

$$K_{pts} = \frac{B_s - (R + a)K_\ell E_s}{P_s(R + a)}$$

Figure 14.2 shows this dependence, but it should be mentioned that precise calculation of B_s for each set of storage parameters requires a special algorithm, which could be considerably simplified by using the design criterion for the introduction of storage.

14.4 Design criteria for the introduction of a storage unit

The optimal solution of the design problem, the so-called generation expansion plan, has to satisfy certain requirements, the principal three of which are

- to generate the amount of energy during a given time as required by the demand side of the power system;
- to be capable of covering the maximal demand; and
- to be flexible enough to cover the minimal demand.

It should be mentioned that the rest of the power system requirements may be calculated on the basis of the above three.

Let us consider the two-step approximation of the load curve given in Figure 14.3, where L_{min}, L_{max}, t_{min}, t_{max} represent the minimal and maximal loads and their durations, respectively. Energy E_l consumed by this load is

$$E_l = L_{min}t_{min} + L_{max}t_{max}$$

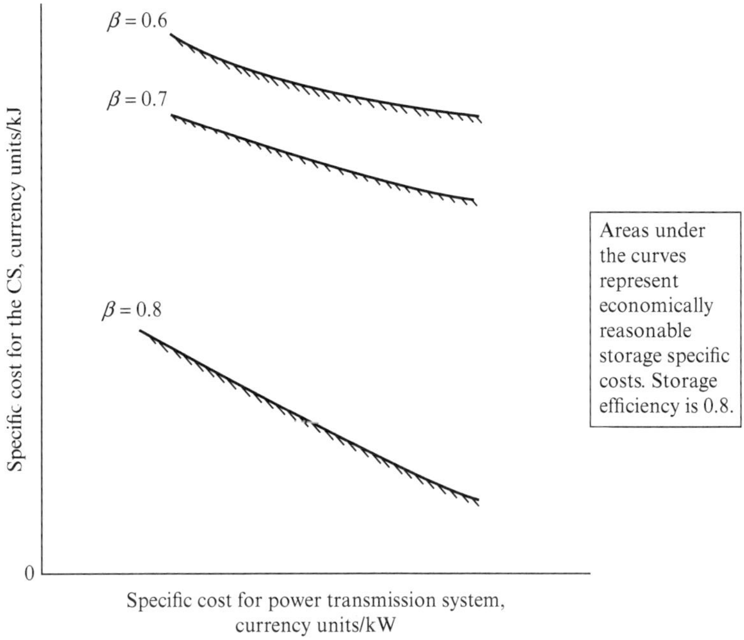

Figure 14.2 Target specific cost for the CS with respect to target specific cost for the PTS for different minimal load factors β

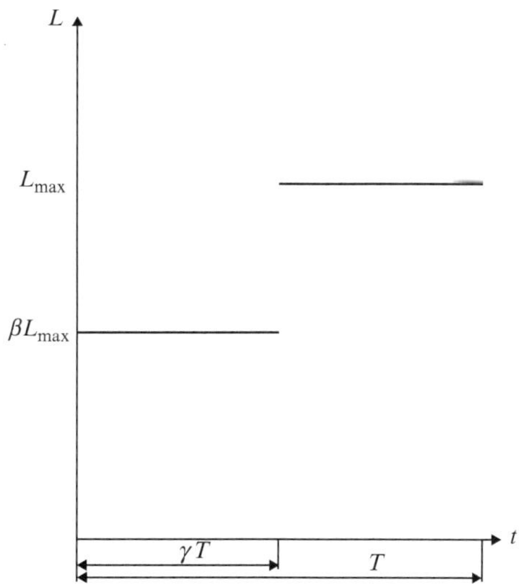

Figure 14.3 Two-step load model

As is clear from Figure 14.3, the load similarity period T (day, week, etc.) consists of t_{min} and t_{max}, so that $t_{min} = T - t_{max}$.

Using the definitions of the load curve parameters, namely minimal load factor

$$\beta = \frac{L_{min}}{L_{max}}$$

and minimal load duration factor

$$\gamma = \frac{t_{min}}{T}$$

it is possible to rewrite the energy equation as follows:

$$\beta L_{max}\gamma T + L_{max}(T - \gamma T) = E_l$$

or

$$L_{max} = \frac{E_l}{[T(1 + \beta\gamma - \gamma)]}$$

It is clear from the latter equation that the three parameters – energy consumed, minimal load factor and minimal load duration factor – can describe completely the two-step load diagram, which contains all the information and could therefore be used as initial data for solving the storage integration design problem instead of using the daily load curve.

For this purpose the simplified load diagram has to satisfy the following requirements:

- The maximal load step L_{max} equals the maximal load of the daily load curve.
- The coefficients β and γ coincide with the minimal load factor and load density factor, respectively.
- Both diagrams cover the same duration of time T, or in other words:

$$E_l = \sum_{i=1}^{n} L_i t_i$$

We are now able to estimate the maximal required rated power capacity for energy storage to perform the load-levelling function. Since, for the load curves of the utilities, $\gamma > \beta$, this will be given by

$$P_s = (1 - \gamma)L_{max}$$

To find out the relevant energy capacity, the value of required discharge time T_d has to be calculated according to the criterion described below.

Let us consider two different expansion plans: with and without the introduction of storage. The cost function for a power system without introduction of storage units will be given by

$$f(x) = RK_0 + U_0$$

The cost function for the power system with a newly introduced storage unit will be given by

$$f(x_i) = RK_i + U_i$$

where K_0, U_0, K_i, U_i = capital and fuel costs, respectively, for the power system without and with introduction of storage units R = discount rate for power equipment.

It is clear that, if the difference between these two functions $\delta f(x) = f(x_i) - f(x_0)$ is positive, then the introduction of a storage unit into a power system will be cost-effective:

$$\delta f(x) = R(k_i - K_0) + U_i - U_0 > 0$$

This allows us to formulate the criterion: unless the first derivative of this difference, with respect to the capacity of a new storage unit N_s, is not negative, it is reasonable to increase the capacity. Hence the equation

$$\frac{d[\delta f(x_i)]}{dN_{si}} = \frac{d[R(K_0 - K_i) + (U_0 - U_I)]}{dN_{si}} = 0$$

or

$$\frac{Rdk_i}{dN_{si}} + \frac{dU_i}{dN_{si}} = 0$$

is the design criterion for introduction of a new storage unit.

Using the two-step approximation of the load curve and the econometric model of the power system with a storage unit, it is possible to rewrite the above criterion as follows: the introduction of energy storage in a conventional power system will be efficient if its turnaround efficiency ξ_s satisfies the non-equation given by

$$\xi_s > \frac{RK_b^* + K_{fb}\gamma T df_f(P_h)/dN_s}{K_{fp}\gamma T df_f(P_p)/dN_s + R(K_p - K_{pts} - K_c t_d)\gamma/(\gamma - 1)}$$

It is clear that the target efficiency, which may be defined as the result of solving the non-equation, is sensitive to a number of parameters such as power sources and storage unit specific costs, fuel costs, shape of fuel consumption curves and variation of load curve parameters. According to the results of sensitivity analysis, storage target efficiency is approximately 75%.

A special algorithm, based on the use of the above-mentioned criterion, allows us to find the optimal storage parameters for introduction into the supply side of power systems using standard power system expansion planning software.

Chapter 15
Effect of energy storage on transient regimes in the power system

15.1 Formulation of the problem

Any power utility suffers from regimes in which power swings take place, the so-called transient regimes. In these regimes undesirable oscillations of frequency and voltage take place, decreasing or removing the utility's ability to transmit generated power to consumers. The problem arises of how to damp these oscillations.

For the purposes of power system analysis, the problem is usually divided into two subproblems that may be solved independently. First, generated power has to be transmitted stably through the existing transmission lines, so there is the subproblem of steady state stability, given the necessity of ensuring the quality of the transient regimes under the small disturbances. This problem occurs with any normal regime and may be considered as a permanent problem in a power utility; it is therefore included in the set of system constraints given in Chapter 14.

Second, the power system has to recover promptly after strong disturbances like short circuits, so the subproblem of transient stability endurance has to be considered carefully in the design stage of power system planning.

This division is reasonable since it takes into account different durations of oscillations and, therefore, allows use of different methods of analysis and a different range of simplifications for the description of a power system.

The problem of damping of oscillations can be solved quite successfully by using so-called power system stabilisers (PSS), static VAr compensators (SVC) and, in the case of strong disturbances, by the application of shunt resistor brakes. Certain types of storage systems with a small time of reverse could also have an effect on solving these subproblems. Active power may be generated or consumed by a storage unit using the deviation of equivalent generator angular velocity as a feedback signal, thus damping the active power oscillations. Using the deviation of voltage as a feedback signal, it is possible to maintain the desired level of voltage by controllable generation or consumption of reactive power. All types of storage except TES may be used for these purposes provided they are equipped with appropriate regulators with properly chosen feedback gains.

The main problem in these cases is to find the essential parameters of the storage unit – its rated power, energy capacity and time of reverse – to ensure damping of oscillations caused by the relevant power swing.

The configuration of the model power system generally used to analyse the effectiveness of the introduction of storage instead of PSS and SVC is the most basic one machine-double circuit transmission line-infinite bus system corresponding in full scale to a 200 km long, 500 kV transmission line connecting a 2 000 MW conventional thermal power station, comprising four 500 MW units, to a large power system. It is not a purely academic example since there are many schemes of such configuration in Russian, Canadian and American utilities.

To investigate the role of energy storage in the transient regimes of the power system, an adequate mathematical model is needed. The model we consider has to comprise elements that describe the transient properties of the electromechanical and electromagnetic parts of the power system.

15.2 Description of the model

The utility to be analysed consists of the following elements:

- equivalent generator, the mechanical input P_m of which is regarded as constant by neglecting the effect of the governor. This supposition, in fact, sets the limitation of storage discharge time – it should be within a few seconds;
- automatic voltage regulator (AVR), represented by the block diagram in Figure 15.1, and the relevant voltage;

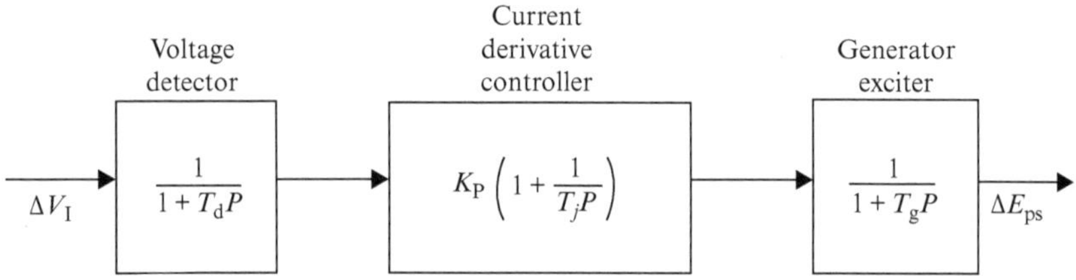

Figure 15.1 Block diagram of an automatic voltage regulator

T_d, T_i, T_E: voltage detector's, derivative controller's and exciter's time constants, respectively

- double chain transmission lines, represented by pi-diagram circuits;
- energy storage unit, represented by its mathematical model, as described in Chapter 2; and
- infinite bus, represented by constant voltage and frequency.

The mathematical model of this utility will be a set of equations describing the performance of each element and their interaction in a system.

The model comprises four equations describing the mechanical and electromagnetic transient processes as well as the active and reactive power balances in the reference utility. They are:

(i) Equivalent generator rotor movement equation, or torque equation:

$$\frac{T_j d^2\delta}{dt^2} + \frac{D_m d\delta}{dt - (P_m - N_{gs})} = 0$$

(ii) Electromagnetic transient processes in the rotor windings – the change in the flux linkage of the field winding equation:

$$\frac{T_{do} dE'_q}{dt + E'_q} - E_{qe} + I_d(x_d - x'_d) = 0$$

(iii) Active and reactive power balance equations in the node with the energy storage unit:

$$N_{gs} - N_{sb} \pm P_s = 0$$
$$Q_{gs} - Q_{sb} - Q_s = 0$$

(iv) Active and reactive power flow between the generator terminal and the storage node, and between this node and the infinite bus, given by the well-known equations:

$$N_{gs} - \frac{E_q V_s \sin(\delta - \delta_s)}{x_1}$$

$$Q_{gs} = \frac{[-V_s^2 + V_s E_q \cos(\delta - \delta_s)]}{x_1}$$

$$N_{sb} = \frac{V_s V_b \sin\delta_s}{x_2}$$

$$Q_{sb} = \frac{(V_s^2 - V_s V_b \cos\delta_s)}{x_2}$$

where T_j = inertia constant of generator
T_{do} = d-axis open circuit transient time constant of the equivalent generator
D_m = damping factor of generator
P_m = mechanical input of generator
$N_{gs}, N_{sb}, Q_{gs}, Q_{sb}$ = active and reactive power flows between generator terminal, storage node and infinite bus, respectively
$\pm P_s, Q_s$ = active and reactive power flow from (to) the storage unit, respectively
δ = torque angle based on the infinite bus
δ_s = phase (angle) of storage node voltage
I_d = d-axis stator current projection
x_d = d-axis synchronous reactance
x'_t = transformer reactance
x_{l1} = reactance of the line between generator and storage

$x_1 = x'_d = x_t + x_{l1}$
$x_2 = x_{l2}$ – reactance of the line between the storage node and infinite bus
V_g = generator terminal voltage
V_s = storage node voltage
V_b = infinite bus voltage
E_q = voltage behind x'_d
V_q = q-axis projection of V_g
$E'_q = V_q + I_d x'_d$
E_{qe} = field winding circuit voltage.

With the help of this model, and applying a reasonable level of simplification, it is possible to analyse the influence of storage on the steady state and transient stability of the power system.

15.3 Steady state stability analysis

Let us consider the schematic diagram of a power system with an energy storage unit, as shown in Figure 15.2.

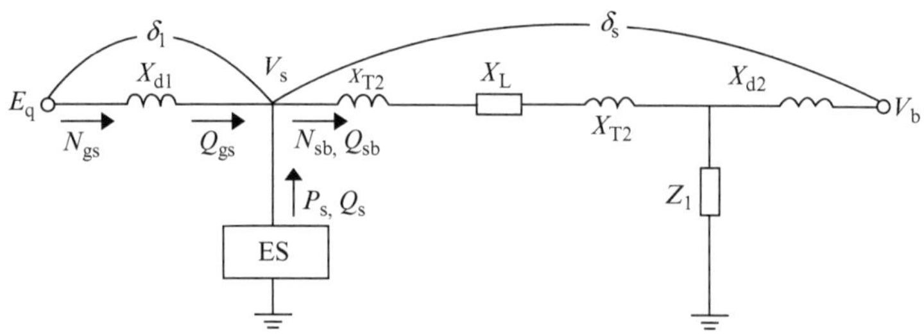

Figure 15.2 Schematic diagram of a simplified power system with energy storage

A storage unit with appropriate parameters, regulator and properly chosen gains has to maintain the steady state stability of the power system under consideration, and maintain constant voltage in the node in which it is located. Suppose we know the whole set of essential parameters and gains. Then, using well-known software, it is possible to find out whether the power system is stable with a storage unit and what the value of the voltage deviation is. By varying the parameters and gains we may find out the marginal parameters of the storage unit and the crucial values of gains to ensure a desirable voltage deviation and the possibility of transmitting generated power through the transmission line stably.

In that case the model can be simplified with the following suppositions:

- Electromagnetic transient processes in the stator circuits, damping circuits and transmission lines are not considered.
- Since the AVR is a highly efficient means for the provision of steady state stability, our particular interest is to investigate the stability of systems in which generators are not equipped with AVR, where the storage unit has to perform the function of that regulator.

Linearising the above-mentioned set of four equations around an operating point, on the assumption that P_m and E_q are constant, we obtain:

(i) Torque equation:

$$[T_j P^2 + dN_{gs}/d(\delta - \delta_s)]\delta(\delta - \delta_s) + T_j P^2 \delta\delta_s + dN_{gs}/dV_s \delta V_s + dN_{gs}/E_q \delta E_q = 0$$

(ii) Equation for change in the flux linkage of the field winding:

$$T_{do} P dE'_q/d(\delta - \delta_s)\delta(\delta - \delta_s) + T_{do} P dE'_q/dV_s \delta V_s + (T_{do} P dE'_q/dE_q + 1)\delta E_q = 0$$

(iii) Active power balance equation:

$$\delta N_{gs} - \delta N_{sb} \pm P_s = 0$$

(iv) Reactive power balance equation:

$$\delta Q_{gs} - \delta Q_{sb} - \delta Q_s = 0$$

where δ denotes the variable representing the deviation from the operating point.

The Jacobian of these equations is given by

$$D(P) = \begin{bmatrix} d_{11} & d_{12} & d_{13} & d_{14} \\ d_{21} & d_{22} & d_{23} & d_{24} \\ d_{31} & d_{32} & d_{33} & d_{34} \\ d_{41} & d_{42} & d_{43} & d_{44} \end{bmatrix}$$

where $d_{11} = T_j P^2 dN_{gs}/d(\delta - \delta_s)$

$$d_{12} = T_j P^2$$

$$d_{13} = dN_{gs}/dV_s$$

$$d_{14} = dN_{gs}/dE_q$$

$$d_{21} = T_{do} PdE'_q/d(\delta - \delta_s)$$

$$d_{22} = 0$$

$$d_{23} = T_{do} PdE'_q/dV_s$$

$$d_{24} = T_{do} PdE'_q/dE_q + 1$$

$$d_{31} = dN_{gs}/d(\delta - \delta_s) \pm dP_s/d(\delta - \delta_s)$$

$$d_{32} = dN_{sb}/d\delta_s \pm dP_s/d\delta_s$$

$$d_{33} = dN_{gs}/dV_s \pm dP_s/dV_s - dN_{sb}/dV_s$$

$$d_{34} = dN_{gs}/dE_q$$

$$d_{41} = dQ_{gs}/d(\delta - \delta_s) - dQ_s/d(\delta - \delta_s)$$

$$d_{42} = -dQ_{sb}/d\delta_s - dQ_s/d\delta_s$$

$$d_{43} = dQ_{gs}/dV_s - dQ_s/dV_s - dQ_{sb}/dV_s$$

$$d_{44} = dQ_{gs}/dE_q.$$

The power system requirements for a storage system, which is assumed to ensure steady state stability and is acting instead of AVR, are given in Table 15.1.

Table 15.1 System requirements for energy storage parameters

Application	**Location**	**Energy capacity (J)**	**Rated power (p.u.)**	**Charge/ discharge time (s)**	**Response time (s)**
Improvement of steady state stability and voltage stabilisation	Between generator's terminal and electrical centre	10^7	0.05	0.02	0.02
Spinning reserve and frequency regulation	Load centre	10^8	0.13	0.36	0.12
Transient stability ensurance and countermeasure against blackout	Generator's terminal	10^8	0.4	0.36	0.12
Load-levelling	Load centre	10^{13}	0.37	2×10^4	120
Multifunction usage	–	10^{13}	0.4	2×10^4	0.12

The significance of introducing a storage unit is that it is capable of controlling active and reactive power simultaneously in order to follow the desired active and reactive power δP_s and δQ_s, which may be varied freely. Let us assume, for simplicity, that the active and reactive power of the storage can be altered instantaneously. This is justified, since the time constants of these power controls are sufficiently small in comparison with the period of the power swing.

The reactive power of the storage is used for constant voltage control of V_s, as in the conventional operation of power system stabilisation by means of SVC, that is δQ_s is generated by

$$\delta Q = -K_v \delta V_s$$

Eliminating δV_s, $\delta\delta_s$ and δQ_s in the linearised mathematical model of a power system, we obtain the system state equations

$$T_s \delta\delta'' + D_m \delta\delta' + a' \delta\delta = b' \delta P_s$$

In order to increase the damping of the power swing in the latter equation, the active power of the storage is controlled in such a way that δP_s is generated by

$$\delta P_s = -K_d \delta\delta'$$

The feedback gains K_d and K_v may be determined by eigenvalue analysis. The eigenvalues of the system state equation, derived from the linearised mathematical model, can be calculated. The eigenvalues corresponding to the power swing modes for various values of K_v and K_d are shown in Figure 15.3, where the storage is located at the generator terminal. (The conjugate pair is omitted in Figure 15.3.) Other eigenvalues have little significance on the discussion of steady state stability.

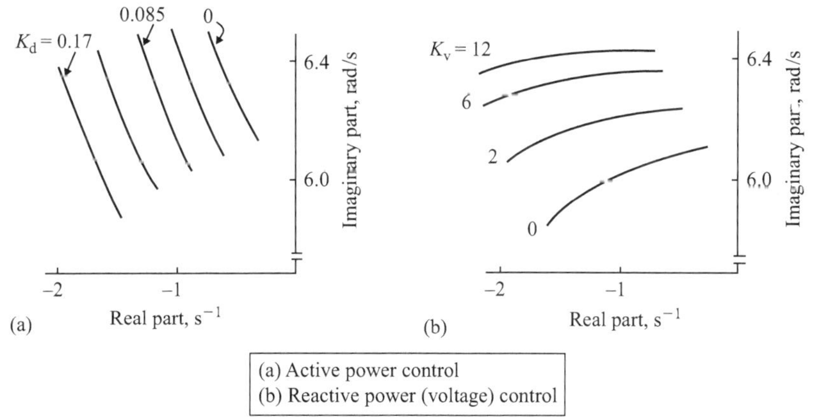

Figure 15.3 Variation of eigenvalues with respect to control gains

The effectiveness of simultaneous control of active and reactive power may be compared with reactive power control by SVC (the case of $K_d = 0$) and with active power control (the case of $K_v = 0$); a comparison is shown in Figure 15.3. In the case of pure reactive power control by SVC, the synchronising power is reinforced by the increase of the gain K_v. This is illustrated by the eigenvalue movement in the upper direction in the complex plane of Figure 15.3. As seen from this figure, the improvement in damping is small.

In the case of pure active control, the damping of the power swing is improved as the gain K_d increases, which is illustrated by the leftward eigenvalue movement in the complex plane of Figure 15.3, while the synchronising power is practically unchanged.

Comparing these results with each other, the synchronising power, as well as the damping, can be reinforced at the same time by choosing K_v and K_d appropriately when active and reactive power control is applied. Variation of the real part of the eigenvalue is shown for different power outputs of equivalent generator in Figure 15.4. The stability effect of simultaneous active and reactive power control by a storage unit may ensure power system steady state stability up to 1.4 p.u., provided the gains K_d and K_v are chosen appropriately.

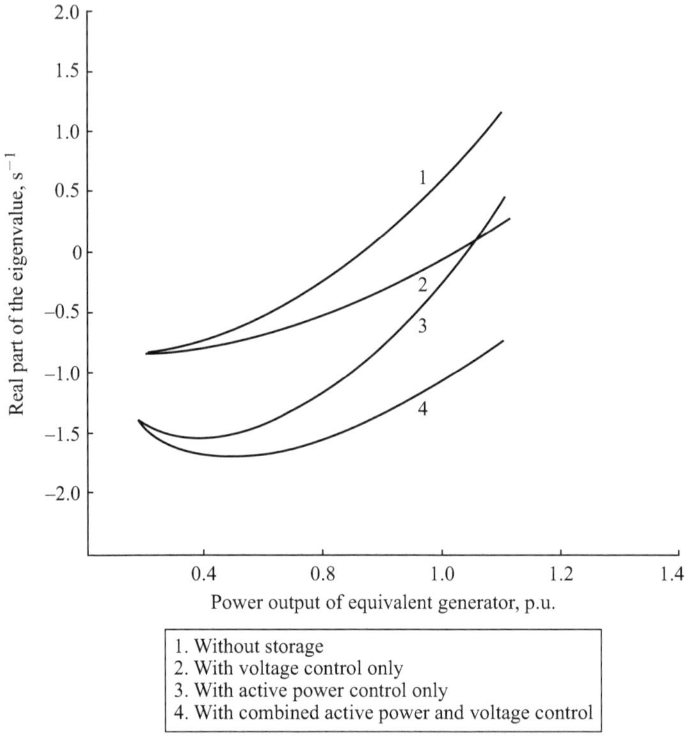

Figure 15.4 Variation of the real part of the eigenvalue for different outputs of equivalent generator

Digital simulations, as well as experimental data already reported, demonstrate that the effect of simultaneous active and reactive power control by storage is more significant than that by SVC, and that transmission throughout may be increased significantly with the proposed stabilising control by storage means.

Varying the rated power of a storage unit, it is possible to find the minimal value that still ensures stability; this value is approximately 5% of the relevant generator's output. The required discharge time may be estimated on the basis of the frequency of irregular power oscillations, which is approximately 1 Hz. The energy capacity of a storage unit may be obtained by multiplying the rated power and discharge time.

15.4 Storage parameters to ensure transient stability

Transient stability is defined as the ability to restore the normal regime after a large disturbance in the power system. A large disturbance is considered to be either a short circuit or any unplanned connection or disconnection of power system elements – sudden loss of load, generators, transformers, transmission lines, etc.

Energy storage can play the same role as the shunt resistor brake so far as transient stability improvement is concerned. But there is a difference – the shunt resistor brake is only able to consume energy, while energy storage has the possibility to work in three regimes: charge (consumption), store and discharge. In addition, the power consumption of shunt resistor brakes depends on the power system voltage and decreases in proportion to the second order of the voltage drop. In contrast, energy storage is able to control the voltage in the node in which it is placed so that active power consumption or delivery can be controlled independently.

Therefore, it is possible to say that the effectiveness of energy storage for increasing transient stability is always more significant than that of a resistor brake of the same size.

Let us consider the effect numerically. Suppose that the energy storage is connected to the power system at the generator terminal, as shown in Figure 15.5, and it is normally in the store mode. The active power $N_{\mathrm{gh}}^{\mathrm{norm}}$ transmitted through the transmission line in this normal regime, as a function of torque angle δ, is given by

$$N_n = \frac{V_{\mathrm{g}} V_{\mathrm{h}} \sin \delta_0}{x_n} = P_0$$

where V_{g} is the voltage at the generator bus
V_{h} is the transmission line voltage (at the infinite bus)
δ_0 is the normal torque angle
x_n is the power system reactance between V_{g} and V_{h}
P_0 is the mechanical power from the turbine.

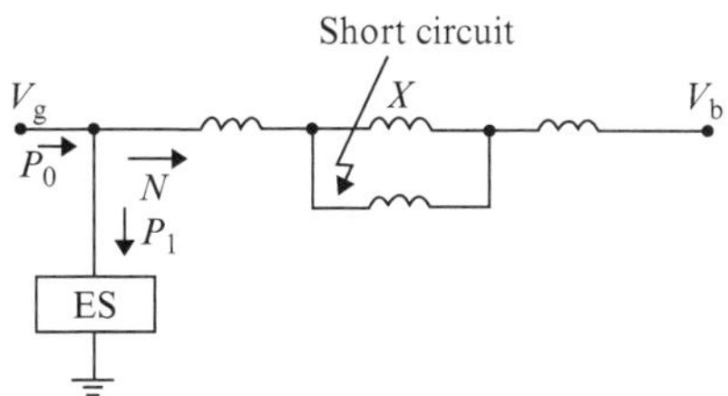

Figure 15.5 Simplified electrical schematic diagram of a power system with energy storage

Let us consider a fault – a 3-phase short circuit in the line between the node where the ES is placed and the infinite bus. In this case no active power can be transmitted through the undamaged chain since all the generated power is diverted through the point of short circuit. The fault is localised and the damaged part is disconnected from the power system in a certain time t_{cutoff}. The undamaged chain of the line is able to transmit less active power than in the normal regime and the transmitted power N after a fault is given by

$$N_{\text{af}} = \frac{V_{\text{g}} V_{\text{b}} \sin\delta}{x_{\text{ud}}}$$

where δ is the changing torque angle after the fault and x_{ud} is the power system reactance including the undamaged part of the transmission line.

At the same time, or even earlier after recognition of the fault in t_{rev}, the energy storage is switched to the charge mode and works in this regime for a period of time t_{c} (see Figure 15.6). The active power balance in that case may be written as follows:

$$N_{\text{af}} + P_{\text{s}} - P_0 = 0$$

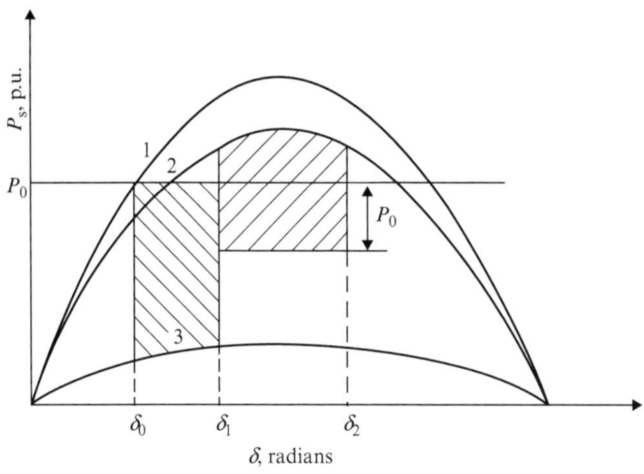

1. Normal regime characteristic
2. Regime after the fault characteristic
3. Fault characteristic

P_0 Turbine power
δ_0 Normal regime angle
δ_1 Cut-off fault angle
δ_2 Storage is disconnected, end of charge regime

Figure 15.6 Generator active power output with respect to torque angle during and after fault regimes (energy storage is charging)

Then after time t_s, when the ES is again in store mode, it is switched to the discharge mode for a period of time t_d (see Figure 15.7). The aim of the storage is to return the equivalent generator's rotor to the stable position with the torque angle δ_4 and with the angular speed equal to 0.

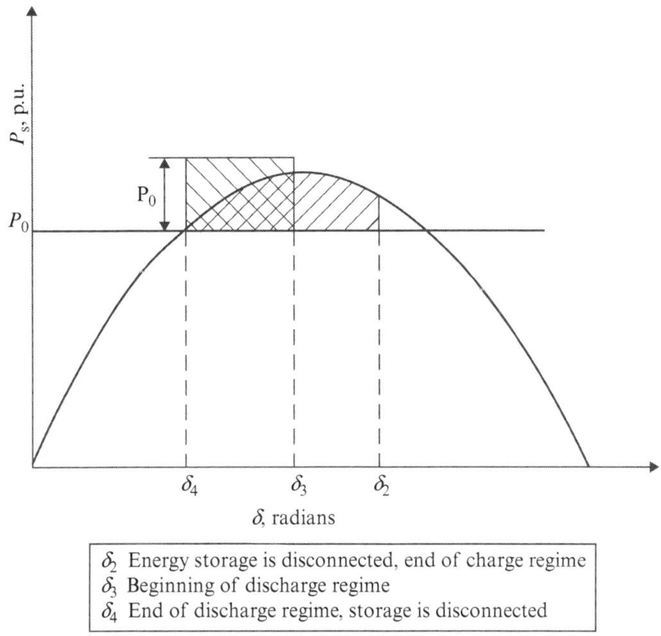

Figure 15.7 Generator active power output with respect to torque angle during and after fault regime (energy storage is discharged and then cut off)

To solve this problem, it is necessary to find the time dependence of torque angle using the power system mathematical model in full without any linearisation. Mathematically, the aim of ES usage means simultaneous satisfaction of the two following conditions:

$$\delta(t_c + t_s + t_d) = \delta_{st}$$

$$\left.\frac{d\delta}{dt}\right|_{t=t_c+t_s+t_d} = 0$$

First, the ES has to consume all the spare energy discharged from the accelerated generator's rotor. Using the square equality rule, it is possible to find the maximal required power capacity of the storage unit

$$P_s = \frac{N_{af}(\cos\delta_{cutoff} - \delta_{cr}) - P_0(\delta_{rev} - \delta_0)}{\delta_{cr} - \delta_{rev}}$$

where δ_{rev} = torque angle corresponding to t_{rev}, the time when the fault was recognised and storage was reversed to the charge mode

δ_{cutoff} = torque angle corresponding to t_{cutoff}, the time when the damaged part was disconnected from the transmission line.

δ_{rev} may coincide with δ_{cutoff} in the case when the storage unit is switched to the charge mode at the same time as the fault is localised; in particular, this is correct for the shunt resistor brake:

$$\delta_{cr} = \pi - arcsin\left(\frac{P_0}{N_{af}}\right) = \text{crucial torque angle}$$

As is clear from this expression, the maximal required value of P_s depends on a number of parameters – the equivalent generator's moment of inertia, torque angle in a normal regime, time of power reverse, fault cut-off time, value of active power, which may be transmitted through the undamaged chain, etc. The relevant functions are shown in Figures 15.8–15.10.

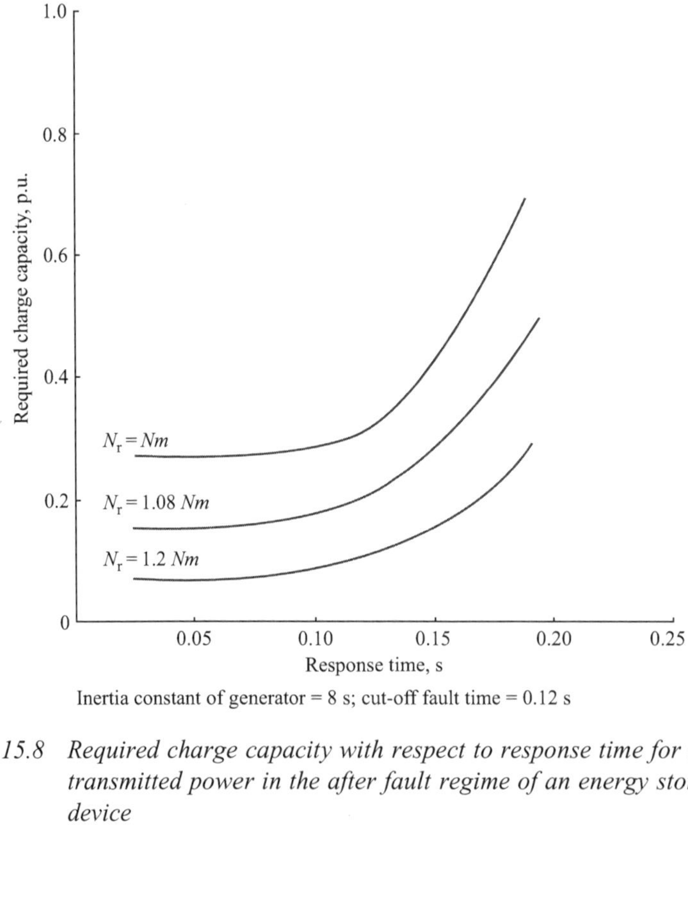

Figure 15.8 Required charge capacity with respect to response time for different transmitted power in the after fault regime of an energy storage device

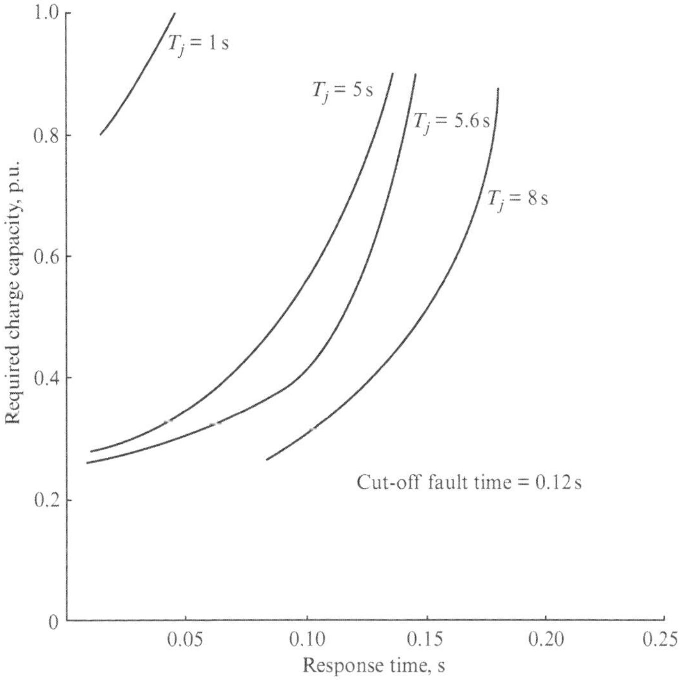

Figure 15.9 Required charge capacity with respect to response time for energy stores with different generator inertia constants T_j

As is clear from Figure 15.8 the desired value of P_s increases as P^{af}_{max} decreases. Since in the 'after the fault' regime the undamaged chain has to transmit all the generated power, that is, $P^{af}_{max} > P_m$, it is possible to define the maximal value of the desired power capacity of storage unit as a function of T_j, t_{rev} and t_{cutoff}.

The value of P_s increases, as is clear from Figure 15.9, with decrease of the moment of inertia T_j. This means that more powerful generators require relatively more powerful storage for provision of stability; their rotors are relatively lighter than those of small generators.

The fault's time of cut-off has a strong influence on the desired P_s; the smaller t_{cutoff} the less P_s is required. A particularly sharp increase in P_s is observed from $t_{cutoff} > 0.16$ s.

As is clear from these figures, the time of power reverse should be no more than the fault cut-off time

$$t_{rev} < t_{cutoff}$$

For the shunt brakes, t_{rev} cannot be less than t_{cutoff}. For fully controlled ES, t_{rev} could be equal to the fault recognition time, which is in the range 0.06–0.1 s. In addition to P_s and t_{rev}, it is necessary to find the required energy capacity E_s

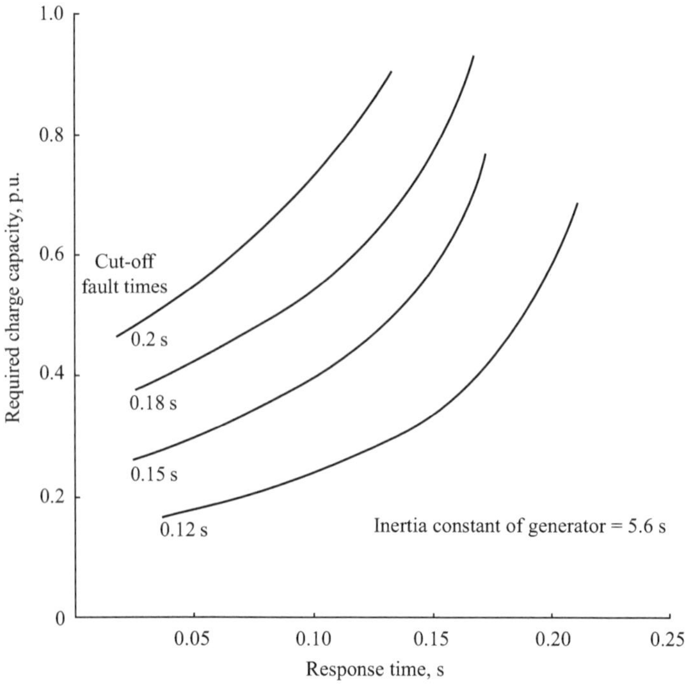

Figure 15.10 Required charge capacity with respect to response time for different cut-off fault times

provided that the storage unit works according to the curves given in Figures 15.6 and Figure 15.7.

Suppose the storage unit's power consumption P_s is constant during the charge and discharge mode, and it is necessary to find out the value of P_s and the energy capacity E, required to ensure stability:

$$E_s = \max\{P_s t_c, P_s td\}$$

The result of solving this problem, given in Table 15.2, is the function $E_s = f(P_s)$, which satisfies the above conditions. The shape of this function also depends on a number of power system parameters: the generator's moment of inertia, fault cut-off time, etc. The relevant functions are shown in Figure 15.11, from which it is seen that there is a point on the curve where the derivative dE_s/dP_s is negative. This means that the storage cost function will have a minimum value. This value will reflect the optimal energy storage parameters, E_s^{opt} and P_s^{opt}, to ensure transient stability.

The system requirements for a storage unit intended for ensuring transient stability are given in Table 15.1.

Table 15.2 Energy storage power and energy capacity required to ensure transient stability for different moments of inertia

T_j(s)P_m (MW) Power Capacity MW			Energy Charged MJ			Energy Discharged MJ		
10	7.5	5	10	7.5	5	10	7.5	5
300	500	800	300	500	800	300	500	800
70	n/s	n/s	78.4	n/s	n/s	8.4	n/s	n/s
80	n/s	n/s	66.4	n/s	n/s	9.4	n/s	n/s
90	90	n/s	60.3	07.1	n/s	10.8	7.2	n/s
100	100	n/s	55.0	87.0	n/s	11.0	7.0	n/s
110	110	n/s	45.1	79.2	n/s	11.0	7.7	n/s
120	120	n/s	46.8	74.4	n/s	10.8	7.2	n/s
130	130	130	48.1	71.5	135.2	11.2	7.8	3.9
140	140	140	49.0	68.6	112.0	11.0	7.0	4.2
150	150	150	51.0	67.5	102.0	10.5	7.5	4.5
160	160	160	51.2	64.0	96.0	11.2	8.0	4.8
170	170	170	52.7	61.2	91.8	10.6	6.8	5.1
180	180	180	54.0	59.4	90.0	10.8	7.2	5.4
190	190	190	55.1	49.4	87.4	11.4	7.6	3.8
200	200	200	58.0	50.0	86.0	10.0	8.0	4.0

n/s means not stable.

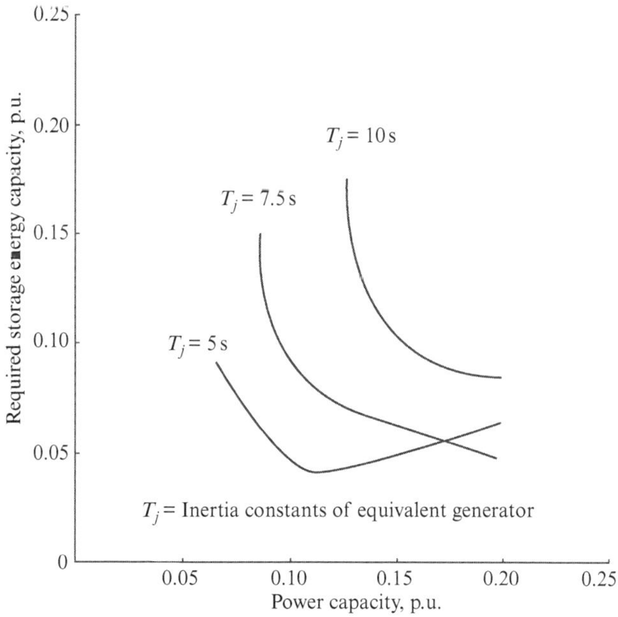

Figure 15.11 Storage energy capacity required to ensure transient stability with respect to power capacity

15.5 Energy storage siting

Generally speaking, there are four possible sites for energy storage in a bulk power system; these were shown schematically in Figure 12.3:

- the generator's terminals;
- load centres;
- intersystem transmission lines; and
- at the consumer's.

Depending on the site of a storage unit, its effectiveness varies with the values of its required parameters. It is therefore possible to formulate a siting criterion as follows: *the difference between the annual benefits of using storage and its annual cost should be maximised:*

$$B_s - (R + a)(K_e E_s + K_{pts} P_s) \rightarrow \max$$

subject to the above-mentioned system constraints.

Control of the stability of a power system is not the only function of an energy storage device, but can be seen as added value to the primary purpose of load-levelling. From the economic point of view, energy storage for load-levelling should be located near to the load centre on the demand side, but it is valuable to evaluate the effectiveness for stabilising control too, for various locations.

The optimal location of storage used for stabilising control can be derived by evaluating the effectiveness of the control of active power and reactive power separately.

As is clear from Section 15.4, improvement of damping by means of active power control is most effective near the generator terminal. Voltage control by

Table 15.3 Evaluation of storage location

	Control	
Location of storage (equivalent distance from the generator) l (km)	**By SVC σ (s^{-1})**	**By combined controlled energy storage σ (s^{-1})**
0	0.5	2.4
50	0.7	2.1
100	0.6	1.4
150	0.4	0.7
200	0.2	0.35

Source: Kovada, Y., Mitani, Y., Tsuji, K. and Murakami, Y.: 'Design of control scheme of SMES for power system stabilization with a facility of voltage control'. Osaka 'Diagaku ChodenkoKogaku Jikken Senta Hokoku (Japan)', vol. 8, 1990, pp. 106–110.

means of reactive power is most effective at the centre of impedance in the transmission system (proved through investigation of the power system-stabilising control using SVC). Considering these effects together, it may be concluded that the region between the generator terminal and the mid-point of the transmission line is a reasonable location for storage in a long-distance bulk power transmission system.

In order to evaluate the effects of location of storage quantitatively, the damping component $\exp(-\sigma t)$ may be calculated, based on the power oscillation mode with a frequency of approximately 1 Hz, which is dominant in the oscillations of torque angle δ. Table 15.3 shows the increment of σ (s^{-1}) in the case without any control. It can be concluded from Table 15.3 that technically the most effective storage location is at the generator's terminal.

It is now possible to bring together information on all the different power system requirements for storage as it performs different functions in a power utility.

15.6 Choosing the parameters of a multifunctional storage unit

Considering the power system requirements for the energy storage parameters given in Chapters 14 and 15, one may conclude that, with appropriate choice of these parameters, the relevant storage unit may be used as a multifunctional device able to solve a wide number of problems in a power utility.

The following order of choice of parameters is proposed for a multifunctional storage unit:

(i) Nominate the desired functions of the storage unit and order them in increase of process duration to find the maximal required discharge time for the storage unit.
(ii) Find out the rated power and relevant energy capacity required for load-levelling and ensuring transient stability, and select the maximal rated power.
(iii) Find the required time of reverse to ensure transient stability, using the rated power selected previously.
(iv) Find the rated energy capacity and the energy capacities required for load-levelling and ensuring transient stability.

It should be mentioned that the most crucial parameter for the use of multifunctional storage is the time of reverse, the desired value of which may be obtained only from flywheels or thyristor convertor-based SMESs, batteries and capacitor banks. The remaining parameters – rated power and energy capacity, and required discharge time – may be obtained from any type of storage equipment provided they are appropriately chosen.

It is reasonable, therefore, to amalgamate different types of storage equipment in one complex, where the parameters of SMES or battery storage, for example, may be chosen for ensuring stability while the parameters of CAES or pumped hydro may be chosen for load-levelling purposes.

The system requirements for a multifunctional storage unit or complex are summarised in Table 15.1.

Chapter 16

Optimising regimes for energy storage in a power system

16.1 Storage regimes in the power system

There are three possibilities for the use of storage in the power system:

- compulsory regime;
- optimal regime; and
- reserve regime.

A compulsory regime arises when the planned load curve coincides with the rated load curve; there is a necessity for the energy storage unit to ensure power balance in the system. It may arise in two different ways:

(i) During the trough period, when load demand is less than the technical minimum of the total installed generation equipment. The energy storage is charged so that stored energy can then be used during peak demand.
(ii) During the peak period, when load demand exceeds the total generation. In this case, the energy stored has to be discharged at its rated power capacity. The necessary energy was accumulated by the ES during the previous load trough.

The situation when the total rated power and energy capacity of the ES are used in a compulsory regime is quite rare, since the storage unit parameters have been chosen under 'cover the winter peak demand and summer trough demand' conditions. It may be considered as an unplanned regime when all the reserve capacity is already used and there is still a shortage of generating power.

An optimal regime arises when part of the rated power and energy capacity of the storage is not used for compulsory load coverage. Usage of the rest of installed capacity of the energy storage unit allows us to change the load on generating units in an optimal way, making it possible to minimise the fuel cost for the energy consumed.

Unused or spare energy storage capacity may also be used as spinning reserve.

Use of the reserve regime provides fuel saving whenever there is a need for spinning reserve. Table 16.1 shows how constant loading decreases the fuel consumption necessary to generate a given amount of energy.

Table 16.1 Fuel consumption (g/kWh)

Type of turbine	Fuel	Rated output	Average loading per year		
		MW	100%	80%	60%
	Coal		339	343	359
K-215-130-1.2	Oil	200	326	330	342
	Gas		322	325	337
	Coal		329	336	350
K-320-240-3	Oil	300	317	325	336
	Gas		313	320	331
K-500-240-4	Coal	500	322	337	355
	Coal		330	336	351
K-800-240-5	Oil	800	316	322	333
	Gas		312	317	328

Nevertheless, the use of storage in an optimal regime provides fuel savings only under certain circumstances – dependent on the type of power source involved, its generation curve and storage efficiency. So, to decide in what regime it is most efficient to use storage, it is necessary to solve the following problem:

$$\begin{aligned} &\text{Maximise} \quad FS_{\mathrm{i}}(P_{\mathrm{sL}}) + FS_{\mathrm{r}}(P_{\mathrm{sr}}) \\ &\text{subject to} \quad P_{\mathrm{sI}} + P_{\mathrm{sr}} + P_{\mathrm{sf}} - P_{\mathrm{s}} = 0 \end{aligned}$$

and the set of system constraints mentioned above,

where FS_{i}, FS_{r} = fuel savings due to optimal and reserve regimes, respectively
P_{sI}, P_{sr}, P_{s} = storage power capacity used in the optimal, reserve and compulsory regimes, respectively.

It is clear that making a decision on whether to use energy storage in a particular regime is an optimisation problem. The main difference between this problem and the energy storage design problem is that, for this stage, all the rated energy storage parameters are already known and are therefore acting as additional limitations to the set of constraints.

If an optimal regime of energy storage is involved, the cost function will be given by the cost of fuel used in the thermal power plants (supply side of the power system) to generate the energy required by the consumers (demand side of the same power system) for a given period of time.

The main purpose of optimal energy storage dispatch planning is to find a regime for the ES for which the fuel cost in the reference system is minimal.

This problem may be formulated as follows:

$$\text{Minimise} \quad K_f = \sum_{j=1}^{m} t_j \sum_{i=1}^{n} F_{cij}(N_{ij})$$

subject to the system and storage constraints above, and

where t_j = duration of the time intervals in which the consumers load curve is approximated by a constant L_i,

n = number of generating units (equal to the number of generating nodes in the reference power system)

$$F_{cij}(N_{ij}) = K_i B_{ij}(N_{ij})$$

where K_i = price of fuel in the ith unit. It should be mentioned that if $K_i = 1$, the fuel cost function becomes the fuel consumption function.

16.2 The optimal regime criterion

Let us consider two steps in the generation curve of a power utility with a storage unit: (i) the peak part with capacity N_p; (ii) the trough part with capacity N_h. In each step there is a certain composition of loaded generation units. As is well known, their loads are determined according to the criterion:

$$\mu = \frac{k_i}{(1 - \sigma_i)}$$

where σ_i = relative increment of the ith power losses
k_i = relative increment of the fuel cost function for the i-th power unit
μ = Lagrangian multiplier.

Let us assume that at each step there is an equivalent generation unit that has loads N_p and N_h and relative increments k_1 and k_2, respectively. To charge the energy storage, let us increase the generation load at the trough step N_h by δN_h, which, in fact, is power charged to storage, $P_{sc} = \delta N_h$.

In that case, it is possible to decrease the generation load at the peak step N_p by δN_p, equalling the power discharged from storage, $\delta N_p = P_{sd}$. The energy accumulated in the storage unit during the charge period t_c, which is the duration of the step N_h, may be given by

$$E_c = N_{sc} t_c$$

The discharge energy may be given by

$$E_d = N_{sd} t_d$$

Following from the mathematical model for energy storage

$$E_d = \xi_c \xi_s(t) \xi_d E_c$$

$$\delta N_p = \frac{\delta N_h \xi_s t_c}{t_d}$$

$$\text{or } N_{sd} = \frac{N_{sc} \xi_s t_c}{t_d}$$

where t_d is time duration of the step N_p and t is the time interval between steps N_h and N_p.

The function for possible fuel saving may be given by the difference in fuel cost between regimes with and without energy storage loading:

$$\begin{aligned}\delta K_f(P_c) &= K_{f2}(N_p)t_d - K_{f2}(N_p - P_d)t_d - [K_{f1}(N_h)t_c - K_{f1}(N_h + P_c)t_c] \\ &= [K_{f2}(N_p) - K_{f2}(N_p - \xi_s P_c t_c)]t_d - [K_{f1}(N_h) - K_{f1}(N_h + P_c)]t_c\end{aligned}$$

It should be mentioned that P_c and P_d are constrained by storage parameter limitations:

$$0 \le P_c \le P_s$$

$$0 \le P_d \le P_s$$

Energy storage usage can be justified economically if the decrease in fuel cost during peak generation (when the ES is discharged) is more than the increase in fuel cost during trough generation (when the ES is charged).

The first derivative of the difference between these fuel costs, with respect to P_s, may be given by

$$\frac{d[\delta K_f(P_s)]}{dP_s} = \left[\frac{\xi_s dK_{f2}(N_p - \xi_s \mathrm{p}_c \mathrm{t}_c/\mathrm{t}_d)}{d(N_p - \xi_s P_c t_c/t_d)} \frac{-dK_{f1}(N_h + P_c)}{d(N_h + P_c)}\right] t_c$$

An extremum condition will be given by

$$\frac{d[\delta K_f(P_s)]}{dP_s} = 0$$

This condition may lead to different results: the value off may reflect maximal or minimal fuel saving due to use of storage. The difference depends on the sign of the second derivative, $d[\delta K_s(P_s)]/dP_s$. In the starting regime, when storage does not participate in load coverage, if this derivative is negative, then use of energy storage will lead to an increase in fuel cost. If the derivative is positive, storage participation is reasonable. Hence, (only) when the first derivative of the fuel cost difference, between peak and trough intervals of the load curve, with respect to energy storage charge capacity is positive, the condition

$$\xi_s \frac{dK_{f2}(N_2, P_s)}{dP_s} - \frac{dK_{f1}(N_1, P_s)}{dP_s} = 0$$

is the optimal energy storage dispatch criterion. This may be used as a basic principle of the algorithm for optimal energy storage dispatch; namely, increase the energy storage charge capacity unless the criterion is not satisfied with given accuracy. The problem is to calculate this derivative numerically.

All the thermal stations in the power system can be represented by an equivalent generating unit with a corresponding fuel cost function. The increment of this function due to energy storage participation in load curve coverage may be given, according to the differential definition, as follows:

$$\delta K_f = \frac{P_s dK_f}{d(N_1 + P_s)}$$

However, this function is also the sum of n terms, which represent n fuel cost functions:

$$\delta K_f = \sum_{i=1}^{n} \frac{\delta N_i dK_f}{dN_i} = \sum_{i=1}^{n} k_i \delta N_i$$

where k_i = relative increment for the ith unit
δN_i – increment generation for the ith unit.

It is possible to use the following simplification with reasonable accuracy:

$$\delta \overline{N}_i = \frac{P_s}{m}$$

where $\delta \overline{N}_i$ = average generation increment of those generation units that apprehend the load changes
m = number of units, $m < n$.

Thus, the increment of the cost Junction may be given by

$$\delta K_f = \delta \overline{N}_i \sum_{i=1}^{m} k_i = \frac{P_s}{m} \sum_{i=1}^{m} k_i$$

and hence we have an equation

$$\frac{dK_f}{d(N_1 + P_s)} = \frac{1}{m} \sum_{i=1}^{m} k_i \quad \text{or} \quad \frac{dk_f}{d(N_1 + P_s)} = \overline{k}$$

where $\overline{k}$ is the average value of the relative increment for those units that apprehend the load changes on the given step of the generation curve.

Then, the practical form of an optimal energy storage dispatch criterion may be given as follows:

$$\overline{k}_2 \xi_s - \overline{k}_1 = 0$$

where $\overline{k}_2, \overline{k}_1$ = average relative increment for the generation units that are additionally loaded due to the participation of storage in the relevant steps of the generation curve.

This criterion may be used as a basis for calculation of the optimal regime algorithm, which uses the calculation of the standard optimal power system regime.

16.3 Criterion for a simplified one-node system

For a simplified one-node system, where grid power losses do not exist and $\sigma_i = 0$, the calculation may be carried out as follows:

$$\gamma_i = K_i$$

and the required derivative may be given by

$$\frac{d[\delta K_f(P_s)]}{dp_s} = -2P_s t_c(A_1 + A_2\xi_s^2 t_c/td) - \frac{dK_{f1}(P_s)}{dP_d} + \frac{\xi_s t_c dK_{f2}(P_s)}{dP_s} = 0$$

The expression for optimal charge capacity may be given by

$$P_s^{opt} = \frac{\xi_s dK_{f2}(P_s)/dP_s - dK_{f1}(P_s)/dP_s}{2(A_1 + A_2\xi_s^2 t_c/t_d)}$$

The second derivative of the function $\delta K_f(P_c)$ is

$$\frac{d^2[\delta K_f(P_s)]}{dP_s^2} = -2t_c(A_1 + A_2\xi_s^2 t_c/t_d)$$

The approximation coefficients A_1, A_2 are positive since the fuel consumption curves are convex. Then, the second derivative is negative and the extremum condition suggests the maximum for the fuel cost function difference. Therefore, $P_c{}^{opt}$ represents the optimal regime for energy storage, that is, the regime in which fuel cost for generated energy is minimal.

It should be mentioned that, according to the definition of the relative increment, it is possible to rewrite the optimal charge expression in the following form:

$$P_c^{opt} = \frac{k_2^{syst}\xi_s - k_1^{syst}}{2(A_1 + A_2\xi_s^2 t_c/t_d)}$$

Only in very rare cases is it possible to calculate in advance the coefficients k_1^{syst} and k_1^{syst} for a given power system, so the expression is not useful for direct calculation. However, it may be used for construction of an algorithm.

Let us suppose that, at the beginning of the calculation, the generation curve steps N_h and N_p have corresponding increments $k_{1(0)}^{syst}$ and $k_{2(0)}^{syst}$ such as

$$\delta = k_{2(0)}^{syst}\xi_s - k_{1(0)}^{syst} > 0$$

Since $\delta \geq 0$, it is efficient to use energy storage. Let us increase the charge capacity by δP_c and the corresponding discharge capacity by $\delta P_d = \delta P_c \xi_s t_c/t_d$, then calculate the new values of system generation capacities $N_{h(1)}$ and $N_{p(1)}$ and corresponding relative increments $k_{1(1)}^{syst}$ and $k_{2(1)}^{syst}$. If, for these new values, $\delta \geq 0$, it is reasonable to increase the charge capacity once more, unless the value of δ becomes, with given accuracy A_c, equal to 0.

Hence expression

$$\delta = k_2^{\text{syst}} \xi_s - k_1^{\text{syst}} - A_c = 0$$

is the optimal storage regime criterion.

This expression may also be used to find the minimal permitted energy storage efficiency under fuel saving conditions, and may also be recommended for use when an expansion planning problem is under consideration. The target efficiency may be given by

$$\xi_s \geq \frac{k_h^{\text{syst}}}{k_p^{\text{syst}}}$$

If the installed energy storage has an efficiency less than the relative increments ratio of the proposed charge and discharge steps, its usage in the optimal regime is unreasonable, and therefore, it should only be introduced in the reserve regime.

The value of minimal storage efficiency depends on the fuel consumption curve of the generation units, their loads and cost of fuel used for base and peak generators.

16.4 Algorithm for the optimal regime

Let us consider an optimal regime for energy storage in a thermal power system where it is used for daily regulation. The given information is

(i) daily load curve stepwise approximation: the load curve is divided into m steps with t_j = duration of each step; usually $m = 24$ h, $t_j = 1$ h;
(ii) for each step the total load demand L_j is given, assumed to be constant during a period t_j;
(iii) structure of generation units and generation curve stepwise approximation (same number of steps and their duration);
(iv) fuel consumption curve and fuel cost for each unit; and
(v) energy storage efficiency, rated power and energy capacity.

It should be mentioned that, if the power system comprises energy storage in addition to conventional sources, the load curve coincides with the generation curve only in one case: when energy storage is not used at all. If energy storage is used these curves do not coincide, and from the economic point of view they represent the different terms in the power system econometric model. The load curve reflects credit for the energy supplied, and the generation curve reflects fuel cost for the energy generated.

If the multistep generation curve is under consideration, the energy storage charge, as well as the discharge, regime may take place in several steps.

During a given period of time, the power system's generating structure may be changed, and therefore, the equivalent characteristics of the relative increments of the fuel cost will also change.

The algorithm comprises six steps:

(i) To start the calculation, it is necessary to compare those two steps of the generation curve with the minimal and maximal relative increments $k_{\rm min}^{\rm syst}$ and $k_{\rm max}^{\rm syst}$. If the optimal regime criterion $\delta < 0$, then usage of energy storage for this particular load curve is unreasonable and the spare capacity has to be used for spinning reserve duty only.

(ii) If the optimal regime criterion is positive, then it is reasonable to increase generation in step with $k_{\rm min}^{\rm syst}$ by $\delta k_{\rm min}$. For a value of

$$k_{\rm min(1)}^{\rm syst} = k_{\rm min}^{\rm syst} + \delta k_{\rm min}^{\rm syst}$$

it is possible to find new generation for all the units that are loaded during a corresponding step. The new load for all generation units, N_j, will be found from the condition

$$\frac{dK_{fi}(N_i)}{dN_i} = \frac{dK_{f2}(N_2)}{dN_2} = \cdots = \frac{dK_{fn}(N_n)}{dN_n}$$

Technically this means that, for ES charge, the most economical generation structure is additionally loaded. The energy storage charge capacity for this step equals the difference between total load and the generation, and has to be compared with its rated power capacity $P_{\rm s}$.

$$P_{ci} = N_j - L_j$$

If $P_{cj} > P_{\rm s}$ then decrease of the $\delta k_{\rm min}^{\rm syst}$ is required. The energy supplied to the storage is

$$E_{cj} = (N_j L_j) t_j$$

and the energy charged to the central store may be given by

$$E_{sj} = (N_j - L_j) t_j \xi_{\rm s}$$

It should be mentioned that $E_{\rm s}$ charge may take place in several steps of the generation curve, so it is necessary to check all the intervals where charge is possible. Let us call these intervals 'possible charge steps'. They can be described as follows:

$$N_j \leq \frac{\left(\sum_{j=1}^{m} L_j t_j\right)}{\left(\sum_{j=1}^{m} t_j\right)}$$

The second step of this algorithm has to be carried out for all the generation curve intervals where

$$k_j^{\text{syst}} < k_{\min(1)}^{\text{syst}}$$

(iii) The energy charged to the central store has to be calculated for all the intervals where energy storage is charging:

$$E_c = \sum_{j=1^m} E_{cj}$$

Then this energy has to be compared with the rated energy capacity E_s, and if $E_c \geq E_s$ it is necessary to decrease $\delta k_{\min}^{\text{syst}}$ and repeat step (ii).

(iv) It is necessary to decrease the maximal relative increment for $\delta k_{\max}^{\text{syst}}$. The new value of relative increment $k_{\max(1)}^{\text{syst}}$ may be given by

$$k_{\max(1)}^{\text{syst}} = k_{\max(0)}^{\text{syst}} - \delta k_{\max}^{\text{syst}}$$

According to the principle of 'relative increments equality', for the new value of $k_{\max(1)}^{\text{syst}}$, it is possible to find the new generation N_j on the corresponding load curve step.

Technically this means the least economic generator units are deloading during peak demand.

The energy storage discharge capacity may be found from

$$P_{dj} = L_j - N_j$$
$$E_{dj} = (L_j - N_j)t_j\xi_d$$

The discharge regime may take place over several intervals too, so it is necessary to check all the steps where discharge is possible. They may be described by

$$N_j > \frac{(\sum_{j=1}^{m} L_j t_j)}{\sum_{j=1}^{m} t_j}$$

Step (iv) has to be carried out for all the intervals where $K_j > K_{\max(1)}$.

(v) Energy discharge from the central store has to be calculated for all the intervals where the storage is discharging:

$$E_d = \sum_{j=1}^{m} E_{dj}$$

Then, this energy has to be compared with the stored energy according to the energy balance:

$$E_d - \xi_s E_c = A_c$$

where A_c is of given accuracy. The value of δk_{max} has to be changed and steps (iv) and (v) have to be repeated or energy balance will not be satisfied.

(vi) The achieved regime has to be checked by the optimal regime criterion:

$$\delta = k_{max(k)}\xi_s - k_{min(k)} - A_c > 0$$

If $\delta > 0$, steps (ii)–(vi) have to be repeated until δ becomes equal to 0.

Figure 16.1 shows the result of an optimal regime calculation for Zagorsk pumped hydro near Moscow.

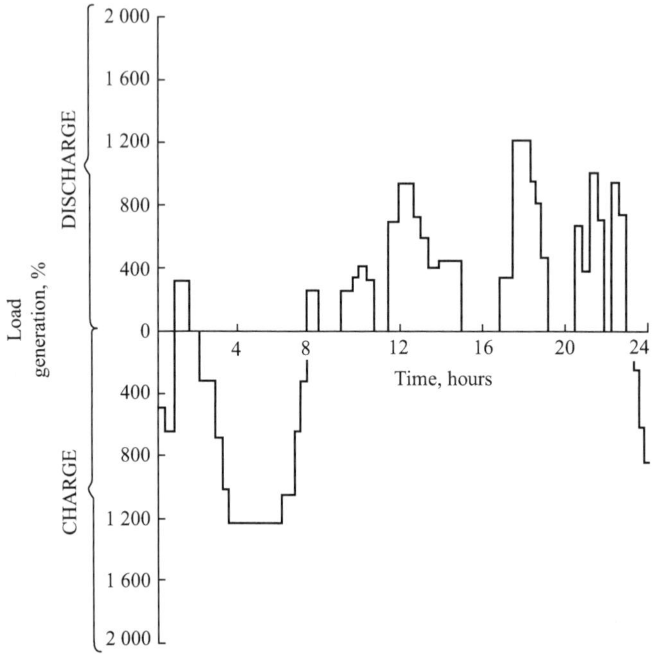

Figure 16.1 Result of storage-independent regime optimisation for a typical winter-day load demand for a typical power utility

Chapter 17

Energy storage and renewable power sources

17.1 Why renewable power sources*

Interest in renewable power sources goes back to 1973. That was a bad year for energy consumers. When the OPEC cartel first became prominent, there was a 3-day working week in the UK and everybody said that oil was running out. Indeed, some said nearly everything would run out in the next 20 years (Club of Rome 1972; Shumacher, 1973). So we needed renewable power sources because that would never run out.

In the event, nothing ran out. Although we were using more and more oil, the reserves increased almost every year for the next 30 years. There was enough coal for hundreds of years and huge reserves of gas were discovered. On top of that, nuclear power capacity increased six times in the next 13 years [until the Chernobyl disaster brought a temporary halt to further developments]. So, by the 1980s, there seemed to be plenty of energy and nobody was particularly concerned about whether it was renewable.

The pessimists had missed the point that there are normally only 20 years of known reserves of any commodity. Prospecting is an expensive business and no company is likely to look for more of anything when there is enough for the next 20 years of production.

Nevertheless, interest in renewable power sources was maintained because another reason arose: global warming. In 1997, the Kyoto Conference was convened. This Conference agreed that the global temperature was rising; that this was due to 'greenhouse' gases, especially CO_2, whose levels in the atmosphere were rising rapidly; and that this increase was due to man-made emissions. One hundred and seventy four countries signed the Kyoto Protocol to that effect but, despite the apparent agreement on climate change being due to man-made emissions, only 34 countries signed Annex 1 to the Protocol agreeing to limit their emissions. Most of the world – 80% of all countries – did not sign it. Most countries had serious doubts about the need to limit their own emissions.

They clearly have no intention of damaging their economies by limiting the operation of their coal-burning plants to reduce their emissions. They regard the effects of climate change as of secondary importance compared with world

* Paragraph 17.1 is written conjointly and *en concordance* with D.T.Swift-Hook and paragraph 17.4 is partly based on his work.

poverty. This was set out clearly in the COP15 Accord produced by the International Conference in Copenhagen in December 2009, which welcomed countries who feel able to limiting their emissions but did not make it mandatory.

Three of the lead countries who produced the Copenhagen Accord are USA, China and India, all heavily reliant upon coal-fired power plant, and, despite their refusals to limit their emissions, these are the countries who held the top rankings for installing wind power in 2008, the most recent complete year before COP15. Between them they installed 61% of the 27 GW of wind power capacity that was installed worldwide. The next year, 2009, they shared an even greater proportion, 65% between them, of the even large amount of wind power installed, 37 GW.

This raises an obvious question: if these countries will not agree to limit their emissions, why are they installing renewable power so enthusiastically, far more than other countries who are desperately trying to limit their emissions? The answer is quite simple: winds and sunshine – the main sources of renewable power – are delivered free everywhere. They do not come from the Gulf like oil or from Siberia like gas. A country does not have to pay foreign currency or fight to protect its energy supplies if they are threatened or the pipelines are shut-off. It could be said that renewable power avoids international conflict. So security of supply is the main reason for the current interest in renewable power worldwide.

The oil spill disaster in the Gulf of Mexico in 2010 reminds us that international disputes can happen between friends as well as foes. President Obama of the USA was pitched against British Petroleum, claiming billions of dollars of compensation. Such oil exploration/production disasters will be avoided if oil is replaced by renewable energy. So avoiding pollution can be claimed as an advantage for the renewables.

Contemporary economies find fuel imports ruinous. It is not such because of the amounts they have to pay. It is because of the uncertainty of fuel prices. From July 2007 the price of oil rose dramatically, doubling to 140 USD per barrel in 6 months. Then, in the next 6 months, it plummeted below 40 USD per barrel. Within another 6 months, the price had doubled again, causing serious disruptions for consumers and producers alike – and countries like UK and Russia are both. Price stability is vital for the health of any contemporary economy.

So there is another big advantage of renewable power. Once it is installed, the energy is free, so renewables are capital intensive. The cost is almost entirely capital expenditure, which is paid for initially and from then on the marginal cost is virtually zero. No price can be more stable than that.

There is an ultimate reason for pursuing renewable power. It is cheap. According to a detailed and authoritative study of generating costs published in 2008 by the California Energy Commission, renewable power in the form of wind is marginally the cheapest. There were several other competing technologies – gas, coal, hydro, geothermal – that all come within 10% or so of wind [while nuclear did not], so although wind was literally the cheapest by a small margin, it would only be safe to say it was *among* the cheapest.

However, there was confidence that wind power would soon become decisively the cheapest. The reason for such confidence was the rate of wind capacity expansion, which is much greater than that of any other type of power-generating plant. It is a well-known principle of production engineering that prices come down in real terms as production capacity increases. When capacity is doubled, the expected savings are at least 10% or 15%. There are savings of scale and a 'learning curve' applies to any industrial production from mobile phones and sewing machines to motor cars and airplanes. There is confidence that this will continue to apply to wind and solar power sources as well.

Wind capacity had been doubling worldwide every 3 years for 15 years or more by 2010 [and in the UK at that point it was nearly doubling every 2 years]. There was no sign of any slow-down, despite the worldwide financial crisis. As this expansion continues, it will inevitably lower the costs of wind power making it the cheapest by an ever wider margin, by 10% or 15% in real terms every 2 or 3 years. Targets and legislation in place in the European Union continue to drive this rate of expansion to 2020 and beyond. For instance, the UK's aims to produce 15% of final energy consumption from renewable sources by 2020, which will require 35% to 40% of the electricity to come from renewable power sources.

The large-scale use of renewable power sources will improve security of supply. It will

- lower dependency on imported fuels;
- produce better balances of trade and payments;
- provide price stability;
- avoid military and political conflicts; and
- reduce pollution levels, including those of CO_2.

Renewable energy in the form of hydro power became cheaper than gas for public lighting as early as 1881 when the world's first public electricity supply system was built for Godalming in Surrey, UK. Hydroelectric power systems rapidly became widespread and associated developments for energy storage, using pumping water, soon followed in Italy and Switzerland in the 1890 s. So the coexistence of renewable sources and energy storage in the electricity generation systems is more than 100 years.

Intermittent renewable power sources in the form of wind or water mills have traditionally been used to pump water by wind mills or to grind cereals, and to saw wood by wind or water mills. The consumption of ground cereals, pumped water and sawn wood does not coincide with their processing (that depends on wind or water flow availability), so technically speaking the storage methods were involved earlier than electricity generation. Water pumping is still a very important application, especially in some rural regions and may be used as a storage method for mini-hydro plants.

Nowadays the general practice of renewable energy sources usage (with the exception of passive solar) is to generate electricity. Ultimately, electrical energy is itself converted into mechanical, thermal, chemical or light energy to provide us with traction and drives, heat and illumination, chemical conversion and storage. So the question arises why energy from intermittent sources should be converted into an intermediate form when ultimately mechanical and thermal energy represent the bulk of demand. The answer to this lies in some exceptional properties of electrical power systems. Some of them are listed below:

- Mechanical–electrical converters are robust, efficient and easily controllable.
- Electrical transmission lines are very efficient and can transport energy over long distances reliably and economically.
- Interconnected electrical power systems are widespread in most of developed countries. Therefore, 'in principle', it is easy to feed the converted energy from intermittent sources into the grid.

It is economically advantageous to offer the energy from intermittent sources in electrical form since it costs more per kWh compared with energy in a solid or liquid fuel. This is because it is 'high grade' energy and, in its generation from fuels, the Carnot cycle imposes a low efficiency of conversion.

17.2 Types of renewable power sources

There are many types of renewable energy, just as there are many types of storage, and most renewable energy comes directly or indirectly as radiation from the Sun. There are the following types of intermittent renewable energy sources:

- wave energy;
- wind energy;
- tidal energy;
- hydroelectric energy; and
- solar thermal technologies and photovoltaics – passive and active solar.

Passive solar uses the Sun's heat directly while active solar converts radiation into electricity, for example, by the photovoltaic effect. Biomass uses photosynthesis to convert sunlight into chemical energy.

Heat from the Sun raises the temperature of the sea surface compared with that at depth and Ocean Thermal Energy Conversion makes use of this temperature difference. Heat from the Sun also evaporates water from the sea [and elsewhere] and the water vapour rises to form high-level clouds from which precipitation provides water, which can drive hydroelectricity schemes.

Differential heating of the atmosphere as hot air rises at the equator and falls at the poles drives stratospheric winds, which are modified by the effects of moving over land and sea and by Coriolis forces as the Earth rotates. Waves are created as winds blow over the surface of the sea.

The gravitational field of the moon drives the Earth's tides and energy can be captured from tidal flows using turbines immersed in water [in the same way that wind turbines capture energy from air flows] or by building barrages to create lagoons that can capture water at high tide to be released at low tide [like a hydroelectric installation].

An exception to all these solar sources of energy is geothermal energy, which is usually treated as renewable, although a thermal borehole loses temperature as heat is extracted, so it has a finite life and it would take a geological timescale to renew that heat.

We do not consider ocean thermal energy conversion (OTEC), geothermal energy conversion (GTEC) and biofuels since these technologies are not intermittent and can be easily accommodated in an electrical power system as any new thermal power plant.

17.2.1 Wave energy

Converting the slow movement of waters to a more easily used and transmitted form, normally electricity, requires special converters. In interaction with waves, these converters must be somehow fixed so that wave forces are resisted. There are several concepts of fixation including

- using seabed for fixing or mooring;
- mounting several converters on a common frame so that relative motion is obtained between them; and
- using the inertial force due to the gyroscopic action of a flywheel relying on the mass and inertia of the device.

Another conversion principle is that of 'overtopping' where wave action causes water to flow over a dam where it is stored, and allowed to run out through a turbine when needed.

Exploiting the low, variable-frequency waves motion, and coupling the power to a fixed frequency and fixed voltage grid system is a complex problem and may be solved in different ways. Various hydraulic and mechanical systems have been proposed, but using air as a working fluid is the most common. Either the air is rectified by valves, or it flows backwards and forwards through a turbine (such as the Wells turbine), which rotates in the same direction, irrespective of the direction of air flow.

Electricity generated in numerous small alternators can be aggregated by a d.c. system. The rapid developments in power electronics have made this technique practicable and economic.

In most cases, wave energy conversion devices are designed to be useful for small isolated consumers who happen to exist by the coastline. Wave conversion devices in this case are installed at or close to the shore. This makes maintenance and power collection much simpler than with devices several kilometres offshore. Since the power needs of communities not connected to the grid are usually quite small, wave energy conversion devices ratings for such purposes could be in the 5–500 kW range. Such devices were installed in Northern Ireland and Norway. As any intermittent power sources installed in an isolated electrical power system they required backup power source (usually a diesel generator) and a storage system to secure energy supply while the diesel is being started.

Wave energy used to be the favourite amongst the intermittent sources especially in the UK but following a very detailed review carried out by the Department of Energy in the early 1980s; it was concluded that wave-generated electricity costs were unlikely to fall to economic levels, though the Norwegian coastline developments claimed much more optimistic costs. Wave power generators are as yet only likely to be useful at suitable coastal sites.

17.2.2 Wind energy

So far, wave power has not been successful in competition with its nearest equivalent, wind power. Wind turbines have been installed at many sites both on the cost and inland. They can in principle be installed offshore. The exploitable resources from the wind around the world are vast. Whenever the mean wind speed exceeds 5 m per second, there are possibilities for exploiting it economically, depending on the costs of competing sources of power. This is not limited to land-based wind turbines. Possibilities exist in installing large wind farms in shallow waters (approximately 30 m depth) around the coasts.

It is necessary to consider here the time structure of wind turbines output. If we look at the spectral analysis of wind speed presented by Van der Hoven back in 1957, we see that wind turbulence has two main frequency domains. There is a high-frequency turbulence (10–1 000 cycles per hour) that causes certain difficulties in maintaining a short-term steadiness of electricity supply. Second, there is a turbulence in times from approximately 5–200 h, corresponding to movements of weather fronts.

Wind turbines or wind mills have been used in rural communities to pump water and to grind cereals for centuries. However, their development had been static for many years. Substantial worldwide interest in wind power was developed in the mid-1970s. There were several reasons for that. The oil prices increases in 1973–74 alongside with great advances in materials and aerospace and computer technologies were among them. This has led to much better understanding of the technology and to the construction of large number of wind farms.

The contemporary development has followed two parallel paths – vertical axis and horizontal axis. The horizontal axis is more familiar and better developed since the traditional wind mills are known for centuries. Both types operate on the principle of aerodynamic lift with the velocity of the blades exceeding that of the wind. In theory, both types can be similar in efficiency but they are fundamentally different in operation modes. The blades of the horizontal axis turbines must be turned into the wind (yawing) whilst the vertical axis wind turbine is omnidirectional. The other fundamental difference is that in the horizontal axis turbine blades rotate in a vertical plane, thus resulting in sinusoidal varying gravitational stresses, whilst the stresses due to the wind remain constant for a given wind speed if we neglect any variation of wind across the circular swept area. Contrary to this the gravitational stresses are constant for vertical axis turbines, and the aerodynamic stresses vary sinusoidally, peaking as the blades cross the wind upstream of the tower. Vertical axis turbines may be developed to larger sizes due to the non-cycling gravitation stresses.

However, for generating electricity, all wind turbines commercially available for use on power systems are nowadays three blade up-wind horizontal axis turbines with tubular towers.

17.2.3 Tidal energy

Tidal energy is in quite a different category to wind. Although it is intermittent, it is more predictable than wind, and is characterised by large quantities being available for a few hours at a time.

Variations in available power take place on a number of timescales. Ebb generation produces electricity for 5–6 h during spring tides, and approximately 3 h during neap tides out of a tidal cycle lasting just less than 12.5 h. A tidal barrage produces two blocks of energy each day, the size and timing of which follow the lunar cycle.

There are two different technologies of extracting energy from the tides. The first is based on the capture of tidal energy as potential energy within a reservoir. The second is based on the direct extraction of the kinetic energy of tidal streams. The latter is a relatively diffuse source and has been developed only at a very small scale, for example, for lighting navigation buoys. So we will describe the concept of tidal power extraction by turbines installed in barrages, which enclose inlets or estuaries and thus develop a substantial differential head.

The technology of tidal power extraction is quite simple: the rising flood tide enters the basin trough sluices (gates openings) and through the turbines idling in reverse (might be used as a source of reactive energy). At high tide, all openings are closed until the tide has ebbed enough to develop a sufficient head across the barrage. The turbines are then opened and generate electricity for several hours, until the head – the difference in level between the emptying reservoir and the next flood tide – has dropped to the minimum at which the turbines can operate. Shortly after, the tide level and reservoir level will be equal and the sluices will be opened. The cycle repeats.

Some projects suggest the exploitation of tidal power plant within a single basin. In this case, the energy output comes in determined by the lunar cycle pulses. There is a possibility for pumped storage incorporation. Turbines in this case might be operated in reverse during the flood tide to raise the level of water in the reservoir to a point higher than could otherwise be achieved. This would result in a higher energy output on the following ebb tide than without pumping.

Other projects employ two basins, one charged on the rising tide and another discharged on the ebb. By installing turbines between the two basins, one has effectively a pumped storage facility incorporated in the tidal power plant capable of compensating the lunar cycle tidal generation to meet load demands.

Pumping can be done during the night-time in the period of low demand when surplus power in the electrical system is most readily available. These occasions during the fortnightly spring–neap cycle usually occur at or about neap tides. This is fortunate, since there is then the greatest scope for increases in water level within the barrage, and the additional power output then occurs during the morning peak demand.

So tidal power differs from wind energy in the scale of installations. With wind energy the sizes of the individual turbines are small – a very positive advantage for placing them near the consumers. With tidal energy, however, the schemes are large – like Severn Barrage and other tidal barrage concepts with a rated output of 7 200 MW. In the event of such projects overrunning on time or/and cost, their economics would rapidly deteriorate.

17.2.4 Small-scale hydroelectric energy

The utilisation of water for mechanical energy goes, as well as wind, back centuries, initially for application such as grinding cereals, and later for a wide range of industrial purposes. It has been used for the direct generation of electricity since 1881. It was hydro power that drove the first electricity power station to supply electricity to the general public at Godalming, Surrey, in 1881.

Hydro power is a renewable source of energy that relies on rainfall and requires terrain with exploitable differences in height or 'head'. Even a few feet heads can be exploited but require large volumes of flowing water – hydro power uses the potential or kinetic energy of water streams. Hydro power plants are usually classified by their rated capacity as follows:

- large – rated capacity 50 MW and above;
- small – rated capacity 5–50 MW;
- mini – rated capacity 500 kW to 5 MW; and
- micro – rated capacity 500 kW and below.

In developed countries, the larger hydro resources have been fully exploited and the attention is now focused on smaller ones. As far as the UK is concerned the remaining potential is small probably less than 2 TWh/year. In less-developed countries, there are still considerable resources yet to be exploited, but the smaller schemes are nevertheless still of interest as possible low-cost, low-technology way to electrification. In the mini- and micro-scale installations, the technological progress is most evident because of the very tight constraint on economics caused by a relatively low energy density.

Throughout the world the annual hydro energy production has risen from 1.8 million GWh in 1986 to 2.3 million GWh in 2009 from both large- and small-scale schemes. Total exploitable capabilities exceed 9.5 million GWh/year. Exploitable capabilities in excess of 100 000 GWh/year are available in Argentina, Brazil, Canada, China, Indonesia, Mexico, Norway, Turkey and Russia. From this list it is clear that hydro potential is not yet depleted.

Although hydro power is, strictly speaking, an intermittent source, it is the easiest to regulate by storage and tends to be regarded as less capricious than the other renewable sources. The resource is widely dispersed and thus less vulnerable to disruption in the event of war or civil disturbances.

17.2.5 Solar thermal technologies and photovoltaics

Near the equator, at the outer atmosphere, our planet receives from the Sun approximately 1 360 J of energy per second over every square metre. That is 1.36 kW of power per square metre. At sea level, we receive approximately 1 kW/m^2, or just over one horsepower per square yard. The energy flux is enormous but it is intermittent. The Sun does not shine at night and sometimes there are clouds in the sky.

The variation of solar radiation itself is shown in Figure 17.1, from which the demand for long-term heat storage is obvious. As soon as solar energy begins to displace non-renewable fuels, the need for efficient energy storage methods will increase.

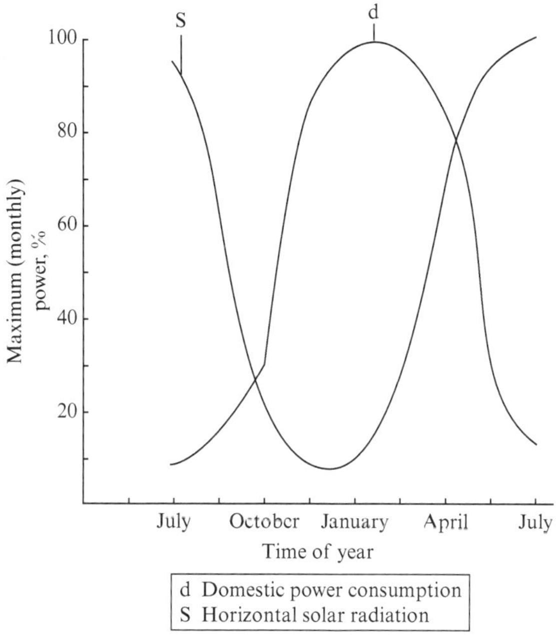

Figure 17.1 Yearly variation of heat demand and solar energy inlet for a northern European house

Solar energy may be concentrated by special collectors and converted into thermal energy and then used for heat supply, water heating, air conditioning, freezing, drying, desalination and in green houses. It may be converted into electricity using thermodynamic cycles. In this case, we can speak about tower-type and modular solar power stations with the thermodynamic cycle. It may also be converted into electricity directly by photovoltaics cascades.

Solar water-heating (SWH) systems have been accepted in most countries for many years. SWH has been widely used in Israel, Australia, Japan, Austria and China. In a 'close-coupled' SWH system the storage tank is horizontally mounted immediately above the roof. No pumping is required as the hot water naturally rises into the tank through flow. In a 'pump-circulated' system, the storage tank is ground or floor mounted and is below the level of the collectors; a circulating pump moves water or heat transfer fluid between the tank and the collectors. SWH systems are designed to deliver the optimum amount of hot water for most of the year. However, even in a tropical environment, there are rainy and cloudy days when the SWH systems may not be sufficient to deliver hot water. So it is advisable to install an SWH system with electrical or gas or other fuel backup. Therefore, many SWH systems (e.g. thermosiphon systems) have an integrated electrical heater in the integrated tank. The installation of an SWH system needs to be complemented with efficient insulation of all the water pipes connecting the collector and the water storage tank, as well as the storage tank (or 'geyser') and the most important warm water outlets. Large SWH systems are economically more efficient than the small ones.

Molten salt can be employed as a heat store to retain heat collected by a solar tower or solar trough so that it can be used to generate electricity when the Sun does not shine. The salt melts at 221 °C (430 °F). It is kept liquid at 288 °C (550 °F) in an insulated 'cold' storage tank. The molten salt is a mixture of 60% sodium nitrate and 40% potassium nitrate, commonly called saltpeter. It is non-flammable and nontoxic, and has already been used in the chemical and metals industries as a heat-transport fluid, so experience with such systems exists in non-solar applications. The liquid salt is pumped through panels in a solar collector where the focused sunlight heats it to 566 °C (1051 °F). It is then sent to a hot storage tank. The latter is well insulated so that the thermal energy can be efficiently stored for up to a week. During the peak demand when electricity is needed, the hot salt is pumped to a conventional steam-generator to produce superheated steam for a turbine/generator similarly to conventional coal, oil or nuclear power plant. A 4-h work of a 100 MW turbine would need tanks of approximately 30 feet (9.1 m) tall and 80 feet (24 m) in diameter. Several parabolic trough power plants are under development in Spain.

There is a possibility to generate electricity using photoeffect phenomenon that occurs in certain semiconductors. Converting into DC requires solar cells usually assembled in photovoltaic cascades. Semiconductor materials presently used for photovoltaics cells include copper indium selenide sulphide. Due to the growing demand for renewable energy sources, the production of solar cells has advanced considerably in recent years.

The efficiency of photovoltaic cells drops with the rise in temperature due to increased electrical resistance. Photovoltaic thermal hybrid solar collectors or hybrid PV/T systems are designed to overcome this disadvantage by carrying the heat away from the PV cells. These systems combine a photovoltaic cascade, which converts solar radiation into electricity, and a solar thermal collector, which captures the remaining energy and removes waste heat from a PV module thereby cooling the cells and thus improving their efficiency by lowering resistance.

Solar photovoltaics generates electricity in more than 100 countries and, while yet comprising a tiny fraction of the 4 800 GW total global power-generating capacity from all sources, is one of the fastest growing power-generation technology in the world. Such installations may be ground-mounted and sometimes integrated with farming or built into the roof or walls of a building. Between 2004 and 2009, grid-connected PV electric system capacity increased at an annual average rate of 60%, to some 21 000 MW, off-grid PV accounts for an additional 3–4 000 MW. Due to the advances in technology and increases in manufacturing scale, the cost of photovoltaic cells has declined steadily since the first solar cells were manufactured. Net metering and financial incentives, such as preferential feed-in tariffs for solar-generated electricity, have supported solar PV installations in many countries. The three leading countries (Germany, Japan and the United States) represent nearly 89% of the total worldwide PV-installed capacity. Germany installed a record 3 800 MW of solar PV in 2009; in contrast, the United States installed approximately 500 MW in 2009. The previous record, 2 600 MW, was set by Spain in 2008.

Photovoltaic power capacity is measured as maximum power output under standardised test conditions (STC) in Watts peak (Wp). The actual power output at a particular point in time may be less than or greater than this rated value, depending on geographical location, time of day, weather conditions, season and other factors. To obtain significant benefit from solar cascades, a seasonal storage would technically be ideal to match the peak output in summer with the peak demand in mid-winter. Solar photovoltaic array factors are typically under 25%, which is comparable with wind turbines.

An indicator of future prospects for the intermittent renewable power sources is the extent to which the sources are actually being used worldwide. At present we have minimal usage of wave energy, some take-up in tidal with wind, hydro and solar leading the field.

17.3 Storage role in isolated power systems with renewable power sources

Small power systems in remote locations pioneered in micro hydro, solar and wind energy applications. In such systems, the cost of transmission line and energy losses argues against connection to a utility grid. Transportation costs, as well as rising fuel costs, argue against the use of fossil fuel generators, such as diesel. Solar and wind systems eliminate these factors, but it is necessary to consider here the time structure of renewable intermittent source (wind turbine, solar array or tidal barrage turbine) output. If we look at the spectral analysis of wind speed presented by Van der Hoven back in 1957, we see that wind turbulence has two main frequency domains. There is a high-frequency turbulence (10–1 000 cycles per hour) that causes great difficulties in maintaining a short-term steadiness of electricity supply. Second, there is turbulence in times from approximately 5–200 h, corresponding to movements of weather fronts. The latter requires a slower control with regard to the operations of the various components of the autonomous electricity system.

So, owing to the intermittency of renewable sources, a certain amount of energy storage capacity is required to match the energy supply and demand over the diurnal cycle.

When we speak about the introduction of renewable sources to an autonomous power system, we are dealing with the introduction of wind power to an existing diesel grid. The original diesel generators may be very much oversized to deal with highly variable loads. Sometimes diesel generator has a capacity two times the peak load to provide a safety margin and room for future expansion of the load. It is not recommended to run diesels for any length of time at small loads. Minimum load of 30% is recommended to avoid technical problems. Even when idling at zero loads the diesel still consumes approximately 30% of maximal fuel consumption. On the other hand, the peak demand may be as high as five times the average load. So an autonomous power grid on its own, without renewable sources, produces rather expensive electricity since it burns too much fuel. And introduction of renewable energy in the form of wind or other sources may decrease the amount of consumed fuel.

Economic wind power installations usually have capacity factors between 0.25 and 0.5 depending on the design of the wind turbines and the energy available in the wind. For the purpose of illustration, let us take 0.33 as a typical capacity factor, equivalent to the wind blowing at its rated speed for one-third of the time and not at all for two-thirds. Such a power system consisting solely of the wind would be short of power for at least two-thirds of the time, a very un-firm sort of power system.

Not only the wind has a highly fluctuating nature, but the power requirements in a small autonomous grid may change very rapidly. Thus, a renewable intermittent source may be fully supplying an autonomous load at one moment and may be in considerable power deficit only seconds later. Hence, the need for a sort of backup (or reserve as it is called in interconnected power grid) is obvious. This results in different combinations of wind turbines, diesel generators, and storage and load control. Recognising that the original diesel generator was grossly oversized, it is possible to use a smaller diesel in combination with a wind turbine and storage. The best solution is a combination of wind-diesel system with long-term storage. It allows diesel start/stop that leads to fuel saving and eliminates load control that sometimes is not suitable for electricity consumers.

Storage as a part of solar water-heating systems is an obvious short-term application if solar energy supplies an appreciable fraction of a household's hot-water demand. The alternative to storage is to have a backup system including not only a conventional water heater but also a delivery system that supplies it with fuel or electricity. The investment a utility makes to provide the ability to meet this occasional energy demand often appears as a 'demand charge' in the consumer's total energy cost: a fixed price that is independent of the quantity of energy actually delivered by the backup water heater. By storing solar-heated water at the site of its use for periods of low solar radiation, the consumer can decrease or even eliminate his/her dependence on external backup systems and reduce the demand charge, with a corresponding net reduction in total energy cost.

Storage of solar-heated water is technically simple and the relevant equipment is available commercially.

17.4 Firm power from intermittent sources

The firmness of a power system is expressed quantitatively by the probability that supply will fail to meet demand despite the spare capacity, that is, the probability that all the power units with spare capacity will fail together. The firmness of an interconnected power system is such that power shortages cause disconnections only a few times in 100 years. A power system can in principle supply any power up to its installed capacity but with different probability of being able to meet different levels of demand.

For almost all the time an interconnected power system operates with spare capacity (different types of reserve approximately 20% of installed capacity) available. It is then the system operator's job to optimise the plant schedule to minimise the cost of generation. The contribution that any individual items of plant make to firm power has little to do with such normal operation. It is concerned with very rare periods in a year when the demand on a power system is maximal: all spare capacity is down for one reason or another and every available unit is called upon to contribute.

No plant has 100% reliability and renewable power source (like wind or solar, for example) has a statistical chance of being available when needed as any other plant does. For instance, nuclear plants in the UK have 60% availability year round, mainly because of planned maintenance shutdowns, but 85% availability during the peak periods that count, so nuclear plants have a capacity credit of 85% of what is installed.

Any intermittent source may on occasion be unable to prevent even a small power shortfall. However, it will sometimes be able to prevent a much larger shortfall, which an equivalent firm energy source could never compensate. If capacities involved are small enough, the larger shortfall is nearly as likely as the lesser and so on average the contribution to system reliability of the intermittent and firm sources is the same. Wind will not be able to contribute so often as other plants but when it can do so, it will contribute correspondingly more. The extra capacity that has to be installed to generate a given annual energy compensates for the lower load factor. And when it is able to generate some conventional plant must somehow stop generation: renewable energy has zero fuel cost. *That has a surprising meaning: conventional plants become intermittent due to renewable sources intermittency.*

The discrepancy between a few outages in a century and one month on average for conventional sources is vast, though slightly better than wind, which is unavailable approximately 8 month per year. This discrepancy does not need to be explained for wind alone but for all individual units of power plants. How then does a power system consisting of all too unreliable components manage to achieve such high levels of availability? The answer is redundancy with diversity. Redundancy may be expressed as spare capacity or reserve, diversity is the hope that failures do not all happen together, that is, they are statistically independent. Wind turbines along with other renewable sources of energy have the considerable merit of improving overall diversity of supply. We have to remember here that unit size of a wind or solar power plant is approximately 50–100 times less than a conventional unit size.

Both the available supply and the maximal demand fluctuate, but by arranging their respective average level to be well separated either by providing adequate reserve capacity, by storage methods or by both, the unfortunate circumstance in which overall demand exceeds supply occurs very rarely. There are strong analogies with safety where many would demand the absolute but where 100% is not in fact attainable and one must be content with a finite, but acceptably, low failure level.

D.T. Swift-Hook pointed out back in 1987 that since power systems are statistical and, as the wind blows some of the time; it can make some contribution to firm power, depending on how much time it is available. He showed analytically that to first order (penetrations up to 20%) wind's firm power contribution is equal to its average power the same way as any other power source. Most of studies indicate that at low penetrations the firm power contribution of intermittent sources is equal to their expected output at time of system risk. This may be applied to all power sources, since the output from any power source conventional or otherwise cannot be guaranteed at all times. Furthermore, the average power that any generator produces throughout a year can alternatively be expressed as total energy per year. Thus, sources of power producing the same annual energy make the same contribution to the firm power on a system since they have the same average power. Sufficient wind turbines to produce the same annual energy as a nuclear or coal power plant will make the same contribution to firm power. It will, of course, be necessary to install more capacity with less availability in order to produce a given annual energy, but that always has to be taken into account when evaluating costs per kWh. The significant point is that the extra capacity needed to provide an equal amount of annual energy or average power from a less-available source is just the right capacity to provide an equal amount of firm power. It must be stressed here that the security of supply of electricity requires continuity of power delivery, not energy.

Extra generating capacity is not needed during the periods of base load; the capacity available only really matters during periods of peak demand. However, extra demand capacity might be needed to absorb excess energy from base-load energy sources (renewables and conventional) instead of shutting down other base-load plants, which is economically unreasonable. And this leads directly to storage methods.

17.5 Storage role in an integrated power system with grid-connected renewable power sources

Any composition of the generating units in the supply side of the power system has to satisfy the main requirement of the demand side: to cover its variable load curve. Renewable energy source output is likewise variable and having zero fuel cost may be treated (*taking no account of transmission lines losses and carrying capacity limitations*) as a negative load. It is often suggested that energy storage will be valuable if intermittent sources, such as wind, wave, solar and tidal, are developed on a large scale. Since output from these sources is variable and, with the exception of tidal energy, subject to practically unpredictable changes in weather, fluctuations in output must be accommodated in the system by regulating the output of conventional plant or by providing storage the same way it is being done to cover the variable power demand. Energy storage is useful because the demand side in a power utility is characterised by hourly, daily and seasonal variations, whereas the installed capacity of the supply side is fixed.

An integrated power system itself, if properly controlled, holds the possibility of acting as a storage device without any additional investment. If there is any change in electricity demand in the power system, it is first accompanied by a slight drop in voltage as energy is extracted from, or supplied to, the electricity grid's equivalent capacitance while equivalent inductance is trying to maintain the current unchanged. So the power system acts as capacitor bank storage but with limited opportunities since there are certain requirements on voltage deviation. The same concept may be applied to the power system's ability to act as a magnetic energy storage device.

The amount of energy stored in the grid's electromagnetic field is substantial but can only be used within tens of milliseconds. After that time, if the change in electricity demand is still occurring, the frequency starts to deviate. This means that energy is extracted from, or supplied to, the rotating parts of the generating system, so the power system acts like flywheel storage. This property of the power system, however, is also limited by the permissible frequency deviation, so one can only use a small part of the energy accumulated in rotating machinery. However, because there are many generators, the capacity of such a flywheel is substantial enough to cover changes in load demand for a few seconds.

The main feature of these kinds of storage is that, being series-connected to the system, they react immediately to any change in load demand, providing an adequate power response.

If the frequency deviation exceeds that permitted by the power system's regulations limit, the steam governor opens or closes the valves and additional energy is extracted from, or supplied to, the enthalpy of the steam in the thermal power station's boilers. There is enough thermal energy stored in the boilers to cover changes in the load demand for a few minutes. So the power system, too, is able to act as a thermal store.

To summarise the above, it is necessary to outline that a power system's ability to act as a capacitor, magnetic, flywheel or thermal energy storage device is effectively built-in, and therefore a 'free' property: generators are playing the role of power transformation systems, while thermal equipment, rotating machinery and transmission lines are playing the role of a central store. The capacities of these stores are limited, however, and therefore the power system's built-in storage can only compensate for short-term fluctuations in load demand. *This property allows accommodating renewable power sources in an electrical power system.*

The situation may be changed considerably if the so-called longitude effect is used. As is well known, there is an hour's time difference for each 1 000 km of longitude. The shape of the daily load curve suggests that, for load-levelling purposes, the beginning of an energy storage charge regime must be from 2 to 8 h away from the required beginning of the discharge regime. If there is a power system comprising two regions, say, one of which has predominantly nuclear or thermal-based generation capacity while the other has predominantly hydro-based generation, and if these regions, being at least 2 000 km apart, are connected by a sufficiently powerful transmission line, it is possible that such a system can have built-in pumped hydro storage, which can be used for load leveling in the interconnected system.

In such a built-in pumped hydro scheme, the transmission lines are acting as waterways in a real pumped hydro, the thermal or nuclear power plants are playing the role of pumps, and hydro plants themselves are acting as a generator while their reservoirs are used like a central store.

The main technical problem to be solved is the necessity to transmit large amounts of power over long distances. The most promising type of line for that purpose is the high-voltage direct current transmission line (HVDC) are being developed in the USA, Canada and Russia. So there is no fundamental technical reason why high proportions of intermittent renewable power cannot be assimilated into the power system.

Contemporary integrated power systems are designed to cope with 'shocks' such as the sudden loss of large thermal power stations and with uncertainties on the demand side. As the tools to deal with these are already available the key question is the extent to which the introduction of large amounts of intermittent energy sources will increase the overall uncertainty in matching supply and demand. This extra uncertainty means that additional short-term reserves are needed to guarantee the security of the power system. These reserves may be placed either on peaking gas turbines or on energy storage units. The costs of additional reserves or storage are one component of the 'costs of renewable power sources variability'. The second is the backup costs and the third constraint costs.

All generating plants make use of a common pool of backup plant that is typically around 20% of the peak demand on the electrical power system. So no special backup provisions need to be made for renewable power. When wind or other renewable power source is introduced, system operators do not rely on the rated power of all the installed renewable sources being available at the times of peak demand, but a lower amount – roughly 30% of the rated capacity at low penetration levels, failing to approximately 15% at high penetration levels. This lower capacity credit gives rise to a modest backup cost. Constraint costs arise when the output from the renewable sources exceeds the demand on the electrical power system. They arise when renewable power contributes more than 20% of demand.

There are numerous technical innovations at various stages of development that can mitigate the costs associated with renewable sources variability. Improved methods of weather forecasting are under development worldwide and could potentially reduce the costs of additional spinning reserve. Most other mitigation measures reduce the costs of managing the electrical power system as a whole. 'Smart grids', for example, cover a range of technology of storage methods inclusive that may reduce the costs of spinning and short-term reserves. Usage of power system itself as a storage along with additional interconnections, including

'Supergrids', also delivers system-wide benefits and aids the assimilation of variable renewable power sources. Electric cars hold out the prospect of reduced emissions for the transport network as a whole and could act as a form of storage on demand side at the consumer's level.

A large power system copes with continuous fluctuations of demand and is operated so as to be able to assimilate these without difficulty. As long as intermittent (solar or wind) energy sources supply only a small fraction of the energy demand in a house, a community or an electricity utility, storage may not be useful since the other components of the energy-delivery system, such as oil- or gas-fired power plants or the utility grid itself, can maintain supply when the Sun is not shining or the wind not blowing. So the perturbations associated with a small proportion (up to 20% of total demand) of intermittent sources would not be noticeable.

In the long term renewable sources will generate increasing amounts of electrical energy. This will lead to capacity remix – the change in optimal thermal plant structure. The primary reason for the optimal plant mix changes is a simple load duration effect: with intermittent sources generating for some of the time, thermal power units will have to supply less energy than would otherwise be the case. Power plants which have higher fuel costs but which are cheap to build therefore become more attractive compared to high capital cost units. This effect may be reinforced by operational factors: high capital cost plants are designed for base-load operations and often incur substantial start-up and part loading costs.

Renewable energy sources might economically provide part of base and intermediate energy demand thus complementing (or competing with) base-load power plants. The intermittent character of this power tends to reinforce the load duration effect of capacity remix. The magnitude and value of capacity remix greatly depends on the costs and characteristics of the power system demand and supply side. Two competing scenarios may be considered: base-load generation (nuclear, large conventional, hydro and renewable) and storage partly replacing peaking generation *or* peaking turbines with renewable power sources steadily replacing conventional base-load plants.

Nuclear capacity, if unconstrained, totally dominates the optimal system at zero penetration of intermittent energy sources. The use of energy storage will possibly displace part of the peaking capacity and will require more base-load capacity (including renewable power sources), which in turn will displace intermediate capacity. So, by installing storage systems for operation with coal and nuclear base-load plants, the utilities will directly support the penetration of intermittent renewable power generation in the longer term. Another scenario assumes substantial changes in power system optimal structure with nuclear power plants steadily being replaced by peaking turbines – they can be quickly and cheaply started or stopped. If the power system is being re-optimised according to this scenario, the capacity of base-load power plants continues to decline with increasing intermittent energy sources penetration rather than leveling off, the amount displaced being not far from removed from the mean winter wind power. (The other source of intermittent energy – solar radiation has less impact in winter.)

The effects of energy storage and/or renewable power source introducing into a power system originally designed without them are not merely financial; all aspects of generation expansion planning need to be considered. Factors related to the cost of this newly introduced unit have to be considered, including an economic evaluation of its performance characteristics. The impact of dispersed energy storage integrated within the system has to be considered, including the effects of distributed units on system stability transmission lines losses and spinning reserve requirements.

At low penetrations of renewable sources no backup is needed to maintain system reliability and storage capacity will not be useful. Intermittent sources-generated electrical energy will normally be fed into utility grids, which have adequate reserve capacities for damping the fluctuations from the intermittent input. The same way they balance their electricity loads.

Interconnected grids can easily accommodate the variability of electric energy from renewable sources as long as their penetration in power systems is not excessive. The situation would change, however, if intermittent energy sources-generated electricity reached a significant fraction of the total power output. Let us consider tidal, wind and solar power penetration.

If the output of the tidal barrage is fed into an interconnected grid system, then the power stations with the highest avoidable running costs, such as gas turbines or older coal-fired plant, can be either shutdown or throttled back in order to absorb the tidal power in a way that maximises its value. The predictability of the output from a tidal barrage means that this absorption can be planned as far ahead as necessary. If the grid system is not sufficiently developed to absorb this predictable but intermittent energy produced each day by a tidal power scheme, which follow the lunar cycle, then a two-way generation, which produces two smaller blocs of energy per tide at around 10–15% greater unit cost, may be preferable. Two-way generation has been chosen for a number of very small schemes in China, which supply local communities that are not connected to the grid. The power system is able to accommodate an un-retimed output from single-basin barrages generating pulses up to approximately 5 GW without any storage, for the same reason as for solar and wind generation. For larger schemes, up to approximately 15 GW, storage is clearly needed, but it is necessary to compare the economics of providing the storage within the tidal plant with having an independent storage scheme.

In the case of wind energy, recent studies confirm that up to 20% of peak demand on the British power system (which is within the immediate reserve margin) could be absorbed without storage intervention although fluctuations on the timescale of seconds to minutes would require an additional regulation duty to be placed on conventional generating plants, and would increase the need for short-term reserve capacity placed, for example, on the storage system. In a large, multi-megawatt power system there will be a number of wind turbines. If they are geographically separated (and they will be separated), then they will provide a useful smoothing of wind power fluctuations we spoke earlier compared with a wind farm placed on one site. On the other hand, a large expansion of offshore wind capacity, will lead to price volatility because generation from these plants is wind dependent rather than electricity demand dependent. Wind generation is cheap and 'green', but some utilities have so much on their system that their conventional plants cannot handle the ramp up or down as the wind levels fall and rise. In addition, the strong level of political support for renewable energy technologies has caused a boom in particular 'hotspots', resulting in transmission bottlenecks. Transmission systems are ageing and, although many new lines are being built, right of way is an issue. System storage and storage methods at the demand side are the natural solution competing with peaking gas turbines. Storage in strategic locations can help to defer new build.

The impact of solar energy generation on the load curve of a utility is shown schematically (see Figure 17.2 for two levels of solar energy contribution to meeting demand). If solar energy could contribute 5% of the utility's power supply, it would help to meet the daytime peak and displace a certain part of the conventional generating capacity. When the solar contribution rises – above 15% – the conventional equipment of a utility's supply side will face a sharp evening peak. In that case, it might be reasonable to introduce storage to level the load on conventional equipment.

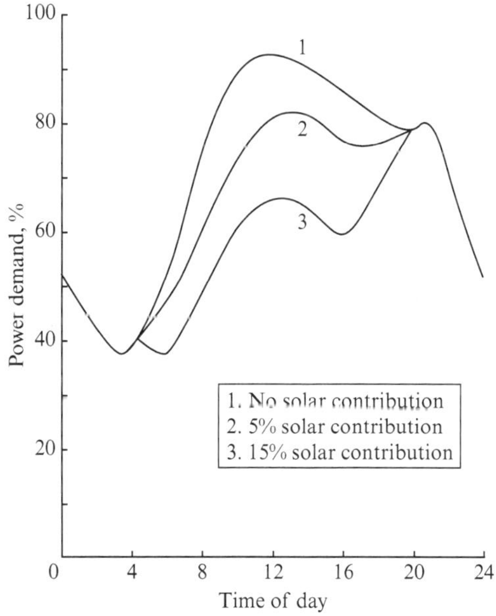

Figure 17.2 Impact of solar energy generation

The Sun does not shine in the evening, so the load demand of the system would be likely to develop a pronounced early-evening peak that would have to be covered by conventional means. Energy storage at the system level could help to flatten this peak, so that the electricity could be generated by constantly loaded, comparatively efficient base-load plants instead of peaking turbines. Solar systems in particular require energy storage because the present high cost of solar cells makes the power system composition, including appropriate storage, economically reasonable. In spite of current uncertainties about optimum system composition and storage capital cost, the onsite storage of solar heat is likely to pay for itself, particularly if the real costs of oil and gas continue to rise.

The addition of energy storage to the power system will certainly lead to a new mix of generation structure. Energy storage will possibly displace part of the peaking capacity and will require more base capacity, which in turn will displace intermediate capacity. So, *by installing storage systems for operation with coal and nuclear base-load plants in the short term, the utilities will directly support the introduction of intermittent renewable power generation in the longer term* (see Figure 17.2).

17.6 Conclusion

Renewable energy sources – tidal, solar and wind particularly – will play a significant role in the supply side structure of future power systems. Their intermittent nature may be smoothed partly by reserve capacity of the power system, partly by finding a special place for them in the load demand curve and partly by using secondary storage. Wind and tidal sources are expected to participate in base generation while solar sources are only suited to the intermediate zone of generation curve.

Renewable sources, including wind, have zero fuel cost and therefore according to the merit order classification would be allowed to generate whenever possible, in preference to thermal plant. At approximately 20% penetration (equal to overall reserve capacity), the variability of the resource can be accommodated and the final savings are a direct result of enabling shutdowns of low merit coal or oil-fired plants. But these shutdowns and subsequent start-ups are costly and affect the lifetime of conventional plant. The introduction of storage systems not only solves this problem, but the base-load generation by nuclear plant would not be affected in case of sufficient capacity storage presence.

The use of any intermittent renewable power source in a power system leads to fuel saving since at the times of renewable generation thermal power plants will partly stop their generation so they do not burn any fuel. Unless the renewables penetration does not exceed power system reserve level these start-stops do not pose any technical problem. But the base-load thermal and nuclear plants will become intermittent, which is not economical and leads to lifetime decrease.

One of the main tasks the energy storage is designed for is load leveling. According to the Electricity Advisory Committee – a panel of experts that advises the US government on industry issues (EAC), the electric power industry in the United States runs at very low capacity factors – as low as 40% Increased levels of renewable energy on the grid will lead to even lower capacity factors for traditional generation sources. So the more intermittent renewable energy sources a power system accepts the more storage facilities will be economically viable. Even if a

Table 17.1 Gives an indication of the energy flows involved

Estimates of worldwide energy flows	
Total solar radiation arriving at the Earth	170 000 TW
Reflected back to outer space	50 000 TW
Heat to Earth (gross solar potential)	80 000 TW
Evaporation (gross hydro potential)	40 000 TW
Wind creation (gross wind and wave potential)	340 TW
Photosynthesis (gross biomass potential)	34 TW
Natural heat outflow from Earth	32 TW
Power in tides	3 TW

storage plant is using electricity from conventional plants in the charge mode, it can reduce emissions of greenhouse gases because it enables all power plants in a system to run at their optimal set point and may therefore be eligible for carbon credits. Storage is a 'greening' technology, because it amplifies existing renewable energy units and allows optimisation of the whole system.

Conclusion

The problem of energy storage integration into a power system is one of the most interesting ones facing power utilities today. In any scenario of power system expansion, there needs to be efficient storage of generated electricity. It is equally essential both for nuclear or coal-fired power plants and for large-scale exploitation of intermittent renewable sources.

The economics of storage plant depends on the mix of other plant on the system: in particular, whether the proportion of large base-load coal-fired or nuclear plants in the system has grown to more than is necessary to cover night-time demand, so that low-cost coal or nuclear-based stored energy becomes available. In this case, storage is complementary to a large nuclear programme but is a valuable part of any large-scale renewable programme. So by installing storage systems for operation with coal and nuclear base-load plants in the short term, the utilities will directly support the introduction of intermittent renewable power generation in the long term.

As power utilities are evolving, large central generating stations are being built progressively further from load centres. Construction of new generation in urban and suburban areas is increasingly limited by environmental concerns and competition for land use. Any form of modular power plant appears to be ideally suited for urban generation, particularly if it is located at established generating stations, as older units are retired, and also at transmission and distribution substations. In particular, energy storage units, which can be charged at night load times and require no external fuel supply, appear to be well suited for installation in overpopulated urban areas.

It is convenient to group all proposed storage duties according to the required duration of their discharge regimes. In the following areas, energy storage will work as a buffer compensating any load fluctuations of several hours duration:

- Utility load-levelling: to improve load factors, reduce pollution in populated urban areas and make better use of available plants and fuels.
- Storage for combined heat and power systems: to improve overall efficiency by offering optimum division between heat and power irrespective of load demands.
- Utilisation of renewable energy in its various forms: to relieve the burden on finite fossil fuel resources and to improve the living environment.
- Storage for remote users.
- Storage for electric vehicles: to replace petrol in the long term, reduce urban air pollution and improve utility plant factors.

The range of several minutes of fluctuation will be covered by the following two means:

- Storage for industrial mobile power units: to provide better working conditions.
- A part of uninterruptible power supply systems: to improve the reliability of supply, especially in confined areas such as warehouses, mines, etc.

The ‘seconds’ range will be represented by diesel-wind generators for output smoothing and storage of necessary energy between pulses of a high-energy particle accelerator.

Millisecond range energy storage units will be used for the improvement of stability, frequency regulation, voltage stabilisation and as a countermeasure against blackout.

Many different types of storage equipment, with very different characteristics, have been developed, and therefore, it is necessary to make a comparison between storage plants and conventional alternatives.

Utilities looking for large-scale energy storage solutions have traditionally turned to pumped storage hydropower. However, such plants are only economic on a large scale and take many years to construct. Available sites for pumped hydro plants are dwindling, both in the United States and Europe. There is up to 50 GW of pumped hydro storage in widespread commercial use in power systems and a further 10 GW is under construction. The main problem with this type of storage is that it is not always easy to find sites suitable for two reservoirs separated by no less than 100 m, not remote from the power grid and having suitable physical characteristics. Massive civil engineering works are required, and since these locations are often in areas of scenic importance, great care has to be taken over the environmental effects of such schemes. For this reason, there are a number of proposals to establish one of the reservoirs hundreds of metres underground.

A conceptually simple way to store energy in a form convenient for power generation is to pump compressed air into an underground reservoir. CAES technology is the only viable bulk storage alternative to pumped hydro. A CAES plant uses off-peak electricity to drive motors that compress air into an underground reservoir. During times of peak demand, the compressed air is withdrawn, heated and passed through an expansion turbine to drive a generator. The world’s first commercial CAES plant was commissioned in 1978 and is located in Huntorf, Germany. Compared with pumped hydro plant, this method has apparent advantages: the air storage cavern can be in either hard rock or salt, providing a wider choice of geological formation, and the density of the energy stored is much higher, that is, there is a smaller minimal size for an economically attractive installation. For a given volume of the underground reservoir, it would be better to store compressed air, since to yield the same power output as a CAES, the reservoir for pumped hydro would have to be very much deeper and could encounter appreciable geothermal temperatures. When the cavern is constructed from salt, it is likely that the cost of extending the storage period of weekly storage will not be as great as with pumped hydro and other schemes.

It is clear that a CAES of comparatively small size (up to 1 million kWh) and short construction time (up to five years) presents a much smaller financial risk to a utility than even the smallest economically reasonable pumped hydro plant of about 10 million kWh. There is, however, one complication: since air gets hot when it is compressed, it must be cooled before it is stored in order to prevent fracture of the rock or creep of the salt. The stored air must then be reheated by burning a certain amount of fuel, as the air is expanded into the turbine that drives the electric generator. There is a need for thermal storage. Against CAES, there is the need to use premium fuels like a distillate oil or natural gas to power the gas turbine. It is possible to overcome this drawback using synthetic fuels – methanol, ethanol or hydrogen instead of natural ones. Methanol has half the volumetric energy density of petrol, is very corrosive and has a high temperature of vaporisation. Ethanol has cold start problems and therefore requires manifold heating. Methanol and ethanol are best used as petrol extenders, but it has been considered that methanol would make an excellent turbine fuel.

Hydrogen generated from water by electrolysis could be stored as a compressed gas, as a liquid or in metal hydride and reconverted into electricity either in a fuel cell or through the conventional gas turbine-generator set. Hydrogen energy storage systems would have considerable flexibility with respect to plant location and operation; its transportation exploits well-established technology for natural gas and, moreover, pipeline transmission over very long distances is cheaper and less objectionable on environmental grounds than electricity distribution.

Today, the bulk of hydrogen is produced from low-cost oil and natural gas and is used almost exclusively for chemical purposes such as synthesis of ammonia, methanol and petrochemicals and for hydrocracking within oil refineries. Hydrogen as fuel amounts to less than 1% of annual production, and therefore, it is difficult at present to define the cost of hydrogen for large-scale applications. It is clear that it cannot at present compete economically with fossil fuels, and this situation is likely to continue until, owing to the current tendency of organic fuel prices to rise, alternative primary energy sources become substantially cheaper than fossil fuels. There are major technical problems to be solved in the production, utilisation and storage of hydrogen but, nevertheless, it is the most promising concept for future environmentally benign power systems.

Thermal energy storage can be either a part of the thermal subsystem of a power system or a secondary source of heat for a consumer. In its first function, TES is not a 'standalone' device but has to be directly linked to the steam generator. Swings up to 50% may be achievable with thermal energy storage designed to raise steam, which is passed through a peaking turbine to avoid overloading the main turbine plant. TESs are ideally suited for load-levelling purposes in the power system.

There are advantages and disadvantages of being an integral part of a thermal power plant: apart from the requirement to be less expensive than peaking capacities, the following practical issues have to be carefully evaluated: plant safety, availability, reliability, flexibility, stability of operation and maintenance.

Various TES concepts have been studied in conjunction with the thermal power station steam cycle, and some of them have been used for many years, but according to modern calculations only pressurised water in lined underground caverns and above-ground oil/rock heat storage in atmosphere pressure vessels (both for steam generation) are found to be economically attractive.

Thermal energy storage is an essential part of a modern CAES concept and is widely known as a source to many as a source of secondary heat at the consumer's premises. It is also very competitive as a store of cold.

The other types of storage equipment – flywheels, chemical batteries, capacitor banks and SMES – have the following advantages in common:

- They are environmentally benign: no requirements for cooling water, no air pollution, minimal noise and moderate siting requirements.
- They have extremely small power reverse time so that power can be delivered or consumed practically on demand, offering increased flexibility in meeting area requirements.
- Power reverse capability can aid in responding to emergency conditions.

Flywheels are under active development, mainly for vehicle applications and for impulse power generation for large-scale storage applications. They offer a number of attractions for energy storage: for short durations of charge–store–discharge cycle they are highly efficient; owing to limitations of materials they only come in relatively small modules, but this drawback for a large power system becomes advantageous for small-scale applications, since they can be made at different sizes and can be installed anywhere they are required in an electricity distribution system; they themselves do not constrain the number and frequency of charge–discharge cycles; and they are environmentally benign. Flywheels are capable of absorbing and releasing energy quickly, but recent studies indicate that even with advanced design they will remain too expensive for large-scale power system applications. Within the power system, application of the flywheel is likely to be in distribution.

Chemical batteries are very attractive for a number of storage applications on both the supply and demand sides of a power system. Batteries have the following excellent properties:

- They store and give up electrical energy.
- Being modular, they can be used flexibly.
- They are largely free of environmental problems.
- They have no mechanical ancillaries.
- They can typically have a short lead time in manufacture.

Containment of reactive materials can be accomplished in the battery itself and corrosion should be eliminated. Thin-film technology used in voltaic pile construction will provide the basis for low-cost production systems.

The modular construction of batteries permits factory assembly, which leads to low site costs and short construction lead times. Since chemical batteries have minimal environmental impact associated with their modest site requirements, high safety and low pollution, and low noise, this allows them to be dispersed optimally in essentially small units close to the consumer, thereby providing a saving in transmission costs compared with other schemes.

Batteries situated close to the consumer are able to smooth the load on the distribution network, thus decreasing the required capacity of substations. They could also be used as an additional source of thermal energy for a district heating scheme, utilising waste battery heat generated during the daily charge–discharge cycle, thus working together with TES at the consumer's premises.

Battery storage for solar electricity is one of the fastest developing areas of application. In most present applications, a rechargeable chemical battery is associated with solar cells. Considerable research and development effort is being made for the creation of a low-cost, high-energy density and reliable batteries for electric vehicles.

Like batteries, capacitor banks have a modular structure and therefore allow factory assembly of standard units, which provides short lead time from planning to installation, thereby reducing capital cost. The main drawback of capacitor bank storage is the low energy density in comparison with batteries, but since capacitors have very small internal resistance, the power density of these storage schemes is very high – they could be used for power multiplication where necessary.

Superconducting magnetic energy storage schemes can directly store electricity with high efficiency, but they are still extremely expensive. SMES seems to have the potential of becoming economically attractive but only on a very large scale. Although this is a high-technology area, there are no insoluble technical problems, but the necessity to work with large units requires sufficient experience that could be gained only by exploitation of small and economically unjustifiable devices. The development and launch costs will therefore be high. (Exploitation of Large Handron Collider near Geneva may help to overcome this problem – it has parameters nearly the same as SMES.)

One may conclude that storage types such as TES, CAES and pumped hydro have a relatively large time of response, and therefore, the possibility of their application is limited compared with flywheels, chemical batteries and SMES, whose time of response is very low. On the other hand, as it is clear that TES, CAES and pumped hydro are suitable for large-scale applications in the power system, flywheels (owing to their limited size) or chemical batteries and capacitor banks (owing to their modular construction) are better suited to comparatively small-scale applications at the supply side of the power system, or, which is very promising, as dispersed storage on the demand side of the power system. Only SMES may be used anywhere in the power system but economic considerations make the prospects of the application of this technically very attractive type of storage rather remote at present.

The idea of combining different types of storage equipment, to make optimum use of their varying properties, holds much promise. The first example should be an adiabatic CAES of which TES is an essential part. Also promising is a combination of CAES and flywheel, or pumped hydro and chemical batteries, in order to have large-scale storage, on the one hand, and quick response, on the other. But all of them have to be justified economically. From the environmental point of view, the CAES concept, which includes hydrogen (generated by renewable energy sources) used as a storage medium and as fuel for the CAES, looks very attractive.

Unfortunately, it is somewhat expensive in the foreseeable future, but fuel price rises could make the concept less pessimistic. The cheapest and simplest of all could be the storage of additional hot water in the boiler circuit of a thermal power plant.

All these devices are artificial secondary storage. Surprisingly, however, the power system itself, if properly controlled, holds the possibility of acting as a storage device without any additional investment.

If there is any change in electricity demand in the power system, it is first accompanied by a slight drop in voltage as energy is extracted from, or supplied to, the electricity grid's equivalent capacitance, while equivalent inductance is trying to maintain the current unchanged. So the power system acts as capacitor bank storage but with limited opportunities since there are certain requirements on voltage deviation. The same concept may be applied to the power system's ability to act as a magnetic energy storage device.

The amount of energy stored in the grid's electromagnetic field is substantial but can only be used within tens of milliseconds. After that time, if the change in electricity demand is still occurring, the frequency starts to deviate. This means that energy is extracted from, or supplied to, the rotating parts of the generating system, so the power system acts like flywheel storage. This property of the power system, however, is also limited by the permissible frequency deviation, so one can only use a small part of the energy accumulated in rotating machinery. However, because there are many generators, the capacity of such a flywheel is substantial enough to cover changes in load demand for a few seconds.

The main feature of these kinds of storage is that, being series-connected to the system, they react immediately to any change in load demand, providing an adequate power response.

If the frequency deviation exceeds that permitted by the power system's regulations limit, the steam governor opens or closes the valves and additional energy is extracted from, or supplied to, the enthalpy of the steam in the thermal power station's boilers. There is enough thermal energy stored in the boilers to cover changes in the load demand for a few minutes. So the power system, too, is able to act as a thermal store.

To summarise the above, it is necessary to outline that a power system's ability to act as a capacitor, magnetic, flywheel or thermal energy storage device is effectively built-in and therefore a 'free' property: generators are playing the role of power transformation systems, while thermal equipment, rotating machinery and transmission lines are playing the role of a central store. The capacities of these stores are limited, however, and therefore the power system's built-in storage can only compensate for short-term fluctuations in load demand. Nevertheless, this very property secures the fact that the exploitation of intermittent renewable energy sources – wind, wave, tidal and solar power – does not present any particular technical problems to large interconnected power systems. Therefore, these energy sources compete on equal terms with conventional sources such as coal, oil and nuclear, provided there is enough installed capacity in spinning reserve – peaking gas turbines in case power transmission network is able to transmit energy generated by these sources. Variability becomes a problem when the proportion of intermittent sources exceeds certain limits – up to 20% of power system installed capacity. And the best solution of this problem is use of different storage technologies or methods.

The situation may be changed considerably if the so-called longitude effect is used. As is well known, there is an hour's time difference for each thousand kilometres of longitude. The shape of the daily load curve suggests that for load-levelling purposes, the beginning of an energy storage charge regime must be 2–8 h away from the required beginning of the discharge regime. If there is a power system comprising two regions, say, one of which has predominantly nuclear or thermal-based generation capacity, while the other has predominantly hydro-based generation, and if these regions, being at least 2 000 km apart, are connected by a sufficiently powerful transmission line, it is possible that such a system can have built-in pumped hydro storage that can be used for load-levelling in the interconnected system.

In such a built-in pumped hydro scheme, the transmission lines are acting as waterways in a real pumped hydro, the thermal or nuclear power plants are playing the role of pumps, and hydro plants themselves are acting as a generator while their reservoirs are used like a central store.

The main technical problem to be solved is the necessity to transmit large amounts of power over long distances. The most promising type of line for that purpose is the high voltage direct current transmission line (HVDC) under development in the USA, Canada and Russia.

There is clearly a limit to the amount of storage plant that is needed to provide daily smoothing. It is clear that the higher the percentage of the total capacity of the generating park allotted to storage, the smaller is the convenience of installing more storage capacity. This is because those means are required to provide longer operating times, with higher costs owing to the larger storage capacities needed, and because the advantages to be obtained from the dynamic services become less important as these services are already provided by the existing means.

The other important aspect is the problem of the choice of the most appropriate characteristics for storage plant; first, the choice of its rated power and energy storage capacity. The ratio between the storable energy and a unit's rated power represents a compromise between specific plant cost, which evidently increases with increasing specific storage capacity, and the number and quality of the services it is able to provide. These qualities improve with increasing specific storage capacity.

In general, it will be convenient for the highest peak services to use methods with specific costs that rise rapidly with storage energy capacity, and to use others for longer durations.

Energy storage usage is being restrained by a lack of regulation to guide utilities and investors. There is no overall strategy for the incorporation of energy storage technique and methods into the power systems, particularly in those with separation of transmission and generation. Energy storage benefits both generation and transmission companies, and this immediately leads to confusion concerning the question of energy storage control. This situation leaves utilities and investors uncertain as to how investments in energy storage should be treated and how costs could be recovered, so neither is willing to invest. The technology should be given more financial and institutional support – alike renewable energy technologies – in order to encourage the construction of storage plant.

Increased levels of renewable energy on the grid will lead to even lower capacity factors for traditional generation sources. One of the main tasks the energy storage is designed for is load-levelling. So the more intermittent renewable energy sources a power system accepts the more storage facilities will be economically viable. Even if a storage plant is using electricity from conventional plants in the charge mode, it can reduce emissions of greenhouse gases, because it enables all power plants in a system to run at their optimal set point and may therefore be eligible for carbon credits. Storage is a 'greening' technology, because it amplifies existing renewable energy units and allows optimisation of the whole system.

The problem of choice of storage parameters appears to be the first to be solved. In order to define the requirements for storage units, power system analysis should be carried out on the following topics:

- Different types of energy storage means in operation at the design stage of the supply side of power utility expansion planning.
- Operating experiences and criteria in electricity power systems with storage plants.

This book should be considered as an introduction to this multidisciplinary problem, solution of which requires considerable work by scientific teams in industry and academia.

With the problems that await the penetration of storage into the supply side structure of the power utilities, it is necessary to start work now.

Only in that way can we have genuine hope for environmentally benign power utilities comprising reasonably distributed 'green' energy sources in the visible future.

Further reading

1. Haydock, J.L.: 'Energy storage and its role in electric power systems'. *Proceedings of World Energy Conference*, Detroit, MI, USA, 1974, pp. 1–28
2. Astahov, YU.N., Venikov, V.A. and Ter-Gazarian, A.G.: 'Energy storage role in power systems'. *Proceedings of Moscow Power Engineering Institute (Moscow)*, 1980, **486**, pp. 65–73
3. Davidson, B.J., *et al.*: 'Large scale electrical energy storage'. *IEE Proceedings (IEEE), Part A*, 1980, **127**(6), pp. 345–385
4. Wood, A.J. and Wollenberg, B.F.: *Power Generation, Operation and Control* (John Wiley & Sons, New York, 1984)
5. Sullivan, R.L.: *Power System Planning* (McGraw-Hill Book Company, New York, 1977)
6. *Electric Generation Expansion Analysis System*, EPRI report EL-2561, vols. 1 and 2, August 1982
7. Kerr, R.H., *et al.*: 'Unit commitment', *IEEE Transactions (IEEE)*, 1966, **PAS-85**, pp. 417–421
8. Ter-Gazarian, A.G. and Zjebit, V.A.: 'Tendencies of power system development in Japan and energy accumulation'. *Enewrgochozyastvo za rubezhom (Moscow)*, 1986, **1** (in Russian)
9. Astahov, YU.N., Venikov, V.A., Ter-Gazarian, A.G., Rebrov, G.N. and Sumin, A.G.: 'Energy storage electrodynamic model'. Author's Certificate No. 1 401 506 USSR, Priority from 26 December 1986
10. Jenkin, F.P.: 'Pumped storage is the cheapest way of meeting peak demand'. *Electrical Review*, 1974, **195**
11. Haydock, J.L. and McCraig, I.W.: *Energy Storage Opportunities in Canadian Electric Utilities Systems* (Acres Consulting Services Ltd., Niagara Falls, ON, Canada, 1979)
12. Astahov, YU.N., Venikov, V.A., Ter-Gazarian, A.G. and Lidorenko, N.S.: 'Energy storage usage in power systems'. *Proceedings of Moscow Power Engineering Institute (Moscow)*, 1984, **41**, pp. 122–128 (in Russian)
13. Astahov, YU.N., Ter-Gazarian, A.G., Boyarintsev, A.F. and Kudinov, YU.A.: 'Electrical power supply system for nuclear plant main pumps'. Author's Certificate No. 1 540 572 USSR, Priority from 9 March 1987
14. Margen, P.H.: 'Thermal energy storage in rock chambers—A complement of nuclear power'. *Proceedings of UN/IAEA International Conference on Peaceful Uses of Atomic Energy*, vol. 4, Geneva, Switzerland, 1971, pp. 177–194

15. Meyer, C.P. and Todd, D.K.: 'Conserving energy with heat storage wells'. *Environmental Science and Technology*, 1973, **7**(6), pp. 512–516
16. Telkes, M.: 'Thermal energy storage', *Proceedings of the 10th IECEC*, Newark, DE, USA, 1975, pp. 111–115
17. Barnstaple, A.G. and Kirby, J.E.: *Underground Thermal Energy Storage*, Final draft report, Ontario Hydro, Toronto, Canada, 1976
18. Collins, R.E. and Davis, K.E.: 'Geothermal storage of solar energy for electric power generation'. *Proceedings of the International Conference on Solar Heating and Cooling*, Miami, FL, USA, 1976
19. Gilli, P.V., Beckmann, G. and Schilling, F.E.: *Thermal Energy Storage Using Prestressed Cast Iron Vessels (PCIV), Final Report*, COO-2886-2, prepared for EDRA, Institution of Thermal Power and Nuclear Engineering, Graz University of Technology, Graz, Austria, 1977
20. Selz, A.: *Variation in Sulphur TES Cost and Performance with Change in TES Temperature Swing* (Energy Conversion Engineering Company, Pittsburgh, PA, USA, 1978)
21. Williams, V.A.: 'Thermal energy storage: Engineering and integrated system'. *Strategies for Reducing Natural Gas, Electric and Oil Costs*, Atlanta, GA, USA, 1990, pp. 277–281
22. Tomlinson, J.J. and Kannberg, L.D.: 'Thermal energy storage'. *Mechanical Engineering* (*USA*), 1990, **112**(9), pp. 68–74
23. Mori, S.: 'Electric power storage and energy storage system: Pumped storage power generation, heat storage and batteries'. *Sho-Enerugi* (*Japan*), 1990, **42**(10), pp. 42–45 (in Japanese)
24. Abdel-Salam, M.S., Aly, S.L., El-Sharkawy, A.I. and Abdel-Rehim, Z.: 'Thermal characteristics of packed bed storage system'. *International Journal of Energy Research* (*UK*), 1991, **15**(1), pp. 19–29
25. Drost, M.K., Somasundaram, S., Brown, D.R. and Antoniak, Z.I.: *Opportunities for Thermal Energy Storage in Electric Utility Applications* (Pacific Northwest Lab., Richland, WA, USA, January 1991)
26. Post, R.F. and Post, S.F.: 'Flywheels'. *Scientific American*, 1973, **229**(6), p. 17
27. 'Economic and technical feasibility study for energy storage'. *Flywheel Technology Symposium Proceedings*, US Department of Energy, Washington, DC, USA, 1978
28. Russel, F.M. and Chew, S.H.: *Kinetic Energy Storage System*, SERC Rutherford Laboratory report RL-80-092, 1980
29. Jayaraman, C.P., Krik, J.A., Anand, D.K. and Anjanappa, M.: 'Rotor dynamics of flywheel energy storage systems'. *Journal of Solar Energy Engineering (USA)*, 1991, **113**(1), pp. 11–18
30. Johnson, B.G., Adler, K.P., Anastas, G.V., Downer, J.R., Eisenhaure, D.B., Goldie, J.H. and Hockney, R.L.: 'Design of a torpedo inertial power storage unit (TIPSU)'. *Proceedings of the 25th Intersociety Energy Conversion Engineering Conference*, New York, NY, USA, 1990, pp. 199–204
31. Thomann, G.: 'The Ronkhausen pumped storage project'. *Water Power*, August 1969, **21**, pp. 289–296

32. Armbruster, T.F.: 'Raccoon Mountain generator-motors'. *Allis-Chalmers Engineering Review*, 1971
33. Baumann, K.M.J.: 'Design concept of the single-shaft three-unit set of Waldeek I pumped storage power station'. *Escher Wyss News*, 1971/72
34. Headland, H.: 'Blaenhau Ffestiniog and other medium-head pumped storage schemes in Great Britain'. *Proceedings of the Institution of the Mechanical Engineers (IME)*, 1971, p. 175
35. Yoshimotoi, T.: 'The Atashika sea water pumped storage project'. *Water Power*, February 1972, **24**, pp. 57–63
36. EPRI Planning Study AF-182, 'Underground pumped storage research priorities', April 1976
37. Mawer, W.T., Buchanan, R.W. and Queen, B.B.: 'Dinorwig pumped storage project—pressure surge investigations'. 2nd International Conference on Pressure Surges, London, 1976
38. Longman, D.: 'Special factors affecting coastal pumped storage schemes'. Far East Conference on Electric Power Supply Industry (CEPSI), Hong Kong, 1978
39. Williams, E.: *Dinorwig: The Electric Mountain* (Pegasus Print and Display Ltd., Leicester, UK)
40. Glendenning, I.: *Compressed Air Storage*, CEGB report R/M/N783, 1975
41. Katz, D.L. and Lady, E.R.: *Compressed Air Storage* (Ulrich's Books, Inc., Ann Arbor, MI, USA, 1976)
42. Herbst, G.E., Hoffeins, H. and Stys, Z.S.: 'Huntorf 290 MW air storage system energy transfer (ASSET) plant design, construction and commissioning'. *Proceedings of the Compressed Air Energy Storage Symposium*, NTIS, Alexandria, VA, USA, 1978
43. Gill, J.D. and Hobson, M.J.: 'Water compensated CAES cavern design'. *Proceedings of the Compressed Air Energy Storage Symposium*, NTIS, Alexandria, VA, USA, 1978
44. Glendenning, I.: *Technical and Economic Assessment of Advanced Compressed Air Storage (ACAS) Concepts*, Electric Power Research Institute report, Project RP1083–1, 1979
45. Stys, Z.S.: 'Air storage system energy transfer plants'. *Proceedings of the IEEE*, 1983, **71**, pp. 1079–1086
46. 'Compressed air energy storage'. *Compressed Air* (*USA*), 1990, **95**(9), pp. 24–31
47. Kimura, H.: 'Compressed air energy storage (CAES) plant', *Kagaku Kogaku* (*Chemical Engineering*) (*Japan*), 1990, **54**(10), pp. 713–716 (in Japanese)
48. Lamarre, L.: 'Alabama Cooperative generates power from air'. *EPRI Journal*, December 1991, pp. 12–19
49. Buchner, H.I., Schmidt-Ihn, E., Kliem, E., Lang, U. and Scheer, U.: *Hydride Storage Devices for Load Leveling in Electrical Power Systems*, CEC report, EUR-7314, 1983
50. Winter, C.J. and Nitsch, J. (eds.): *Hydrogen as an Energy Carrier* (Springer-Verlag, Berlin, Germany, 1988)

51. Salzano, F.J. (ed.): *Hydrogen Energy Assessment*, Brookhaven National Laboratory report BNL 50807, 1977
52. Gretz, J., Baselt, J.P., Ullmann, O. and Wendt, H.: 'The 100 MW Euro-Quebec hydro-hydrogen pilot project'. *International Journal of Hydrogen Energy*, 1990, **15**(6), p. 419
53. Paynter, R.I.H., Lipman, N.H. and Foster, I.E.: *The Potential of Hydrogen and Electricity Production from Wind Energy*, Energy Research Unit report, SERC Rutherford Appleton Laboratory, UK, September 1991
54. Hart, A.B. and Webb, A.H.: *Electrical Batteries for Bulk Energy Storage*, Central Electricity Research Laboratories report RD/L/R 1902, 1975
55. Fischer, W., Hoar, B., Hartmann, B., Meinhold, H. and Weddigen, G.: 'Sodium/sulphur batteries for peak power generation'. *Proceedings of the 14th Intersociety Energy Conference*, Boston, MA, USA, August 1979
56. Talbot, J.R.W.: 'The potential of electrochemical batteries for bulk energy storage in the CEGB system'. *Proceedings of the International Conference on Energy Storage*, Brighton, BHRA, Bedford, 1981
57. Haskins, H.J. and Halbach, C.R.: 'Sodium-sulfur load levelling battery system'. 15th Intersociety Energy Conversion Engineering Conference, Seattle, WA, USA, 18–22 August 1980
58. Bridges, D.W. and Minck, R.W.: 'Evaluation of small sodium-sulfur batteries for load-levelling'. 16th Intersociety Energy Conversion Engineering Conference, Atlanta, GA, USA, 9–14 August 1981
59. Vissers, D.R., Bloom, I.D., Hash, M.C., Redery, L., Hammer, C.L., Dees, D.W., *et al.*: *Development of High Performance Sodium/Metal Chloride Cells* (Argonne National Lab., Argonne, IL, USA, 1990)
60. Tortora, C.: 'New technology for lead-acid batteries'. *Proceedings of 11th IEEE International Telecommunications Energy Conference*, Piscataway, NJ, USA, 1989, p. 66
61. O'Callaghan, W.B., Fitzpatrick, N.P. and Peters, K.: 'The aluminium-air reserve battery: A power supply for prolonged emergencies'. *Proceedings of the 11th IEEE International Telecommunications Energy Conference, IEEE*, Piscataway, NJ, USA, 1989, p. 18.3
62. Abraham, K.M.: 'Rechargeable lithium batteries: An overview', Electrochemical Society 1989 Fall Meeting (Abstracts), The Electrochemical Society, Pennington, NJ, USA, 1989, pp. 58–59
63. Bronoel, G., Millot, A., Rouget, R. and Tassin, N.: 'Ni-Zn battery for electric vehicle applications', Electrochemical Society 1989 Fall Meeting (Abstracts), The Electrochemical Society, Pennington, NJ, USA, 1990, p. 1
64. Jono, M.: 'Characteristics of specialist accumulator batteries for electricity generating systems using sunlight (PS-TL type and CSL type) and some practical examples'. *International Journal of Solar Energy (UK)*, 1990, **8**(3), pp. 161–172
65. Anderman, M., Benczur-Urmossy, G. and Haschka, F.: 'Prismatic scaled Ni-Cd battery for aircraft power'. *Proceedings of the 25th Intersociety Energy Conversion Engineering Conference*, USA, 1990, pp. 143–148

66. Takashima, K., Ishimaru, F., Kunimoto, A., Kagawa, H., Matsui, K., Nomura, E., *et al.*: 'A plan for a 1 MW/8 MWh sodium-sulfur battery energy storage plant'. *Proceedings of the 25th Intersociety Energy Conversion Engineering Conference*, USA, 1990, pp. 367–371
67. Akhil, A. and Landgrebe, A.: *Advanced Lead-Acid Batteries for Utility Application* (Sandia National Labs., Albuquerque, NM, USA, 1991)
68. Olivier, D. and Andrews, S.: *Energy Storage Systems* (Maclean Hunter Business Studies, Barnet, UK, 1989)
69. Gonsalves, V.C.: *Studies on the Sodium-Sulphur Battery* (Southampton University, UK, September 1988)
70. Hart, A.B. and Webb, A.H.: *Electrochemical Batteries for Bulk Energy Storage*, CEGB report RD/L/R 1902, 1975
71. *Engineering Study of a 20 MW Lead-Acid Battery Energy Storage Demonstration Plant*, Bechtel Corporation report CONS/1205–1, San Francisco, CA, USA, 1976
72. Derive, C., Godin, P. and Saumon, D.: 'The possibilities for the development of electrochemical energy storage in the French electricity system'. Commision Economique pour l'Europe Comite de l'Energie Electrique, Rome, Italy, October 1977
73. Bucci, G.D. and Montalesti, P.: 'Ni-Zn battery application to the hybrid vehicle'. *Proceedings of the 12th International Power Sources Symposium*, Brighton, UK, September 1980, Paper 23
74. Godin, P.: 'Batteries for storage in utility networks'. *Proceedings of the CIGRE International Conference on Large High Voltage Electric Systems*, Paris, France, August 1980, Paper 41–06
75. *Development of the Zinc-Chloride Battery for Utility Applications*, Energy Development Associates, EPRI report EM-1417, Palo Alto, CA, USA, 1980
76. Mitaff, S.P.: *Development of Advanced Batteries for Utility Application*, EPRI report EM-1341, Palo Alto, CA, USA, 1980
77. Askew, B.A., Dand, P.V., Eaton, L.W., Obszanski, T.W. and Chaney, E.J.: 'The development of the lithium-metal sulphide battery system for electric vehicle applications'. *Proceedings of the 12th International Power Sources Symposium*, Brighton, UK, September 1980, Paper 22
78. *Sodium-Sulfur Battery Development: Commercialization Planning*, EPRI report GS-7184, Palo Alto, CA, USA, March 1991
79. Soileau, R.D.: 'Testing large storage battery banks'. *Proceedings of the 1990 23rd Annual Frontiers of Power Conference*, Stillwater, OK, USA, 1990, pp. 5.1–5.4
80. Cook, G.M., Spindler, W.C. and Grete, G.: 'Overview of battery power regulation and storage'. *IEEE Transactions on Energy Conversion (USA)*, 1991, **6**(1), pp. 204–211
81. Guk, I.P., Silkov, A.A. and Korolkov, V.L.: 'Extraction of energy from molecular storage'. *Electrichestvo*, 1991, **12**, pp. 53–55

82. Antoniuk, O.A., Baltashov, A.M. and Bobikov, V.E.: 'Calculation of the inductance of the flat buses in capacitance energy storage'. *Electrichestvo*, 1991, **9**, pp. 69–74
83. Ivanov, A.M. and Gerasimov, A.F.: 'Molecular storage of electric energy based on the double electric layer'. *Electrichestvo*, 1991, **8**, pp. 16–19
84. Boom, R.W., Hilal, M.A., Moses, R.W., McIntosh, G.E., Peterson, H.A. and Hassenzahl, W.V.: 'Will superconducting magnetic energy storage be used on electric utility systems?' *IEEE Transactions*, 1975, **MAG-11**, pp. 482–488
85. Peterson, H.A., Mohan, N. and Boom, R.W.: 'Superconductive energy storage inductor-converter units for power systems', *IEEE Transactions*, 1975, **PAS-94**, pp. 1337–1348
86. *Wisconsin Superconductive Energy Storage Project* (University of Wisconsin, Madison, WI, USA, vol. 1, 1974; vol. 2, 1976)
87. Hassenzahl, W.V. and Boenig, H.J.: 'Superconducting magnetic energy storage'. World Electrotechnical Congress, Moscow, 1977, Paper 88
88. Rogers, J.D., Boenig, H.J., Bronson, J.C., Colyer, D.B., Hasselzahl, W.V., Turner, R.D. and Schermer, R.I.: '30 MJ superconducting magnetic energy storage (SMES) unit for stabilising an electric transmission system'. *Proceedings of the Applied Superconductivity Conference*, Pittsburgh, PA, USA, 1978, Paper MB-1
89. Boom, R.W.: 'Superconducting diurnal energy storage studies'. *Proceedings of the 1978 Mechanical and Magnetic Energy Storage Contractors, Review Meeting*, Luray, VA, USA, 1978
90. van Sciver, S.W. and Boom, R.W.: 'Component development for large magnetic storage units'. *Proceedings of the 1979 Mechanical and Magnetic Energy Storage Contractors, Review Meeting*, Washington, DC, USA, 1979
91. Shintomi, T., Masuda, M., Ishikawa, T., Akita, S., Tanaka, T. and Kaminosono, H.: 'Experimental study of power system stabilisation by superconducting magnetic energy storage', *IEEE Transactions*, 1983, **MAG-19**, p. 350
92. Boenig, H.J. and Hauer, J.F.: 'Commissioning tests of the Bonneville Power Administration 30 MJ superconducting magnetic energy storage unit', *IEEE Transactions*, 1985, **PAS-104**, p. 302
93. Mitani, Y., Murakami, Y. and Tsuji, K.L.: 'Experimental study on stabilisation of model power transmission system using four quadrants active and reactive power control by SMES'. 1986 Applied Superconductivity Conference (ASC 86), Baltimore, MD, USA, 1986
94. Masuda, M.: 'The conceptual design and economic evaluation of utility scale SMES'. *Proceedings of the 21st Intersociety Energy Conversion Engineering Conference*, San Diego, CA, USA, 1986, pp. 908–914
95. Shoenung, S.M. and Hassenzahl, W.V.: 'US program to develop superconducting magnetic energy storage', *Proceedings of the 23rd IECEC*, Denver, CO, USA, August 1988, p. 537
96. Andrianov, V.V., *et al.*: 'An experimental 100 MJ SMES facility (SEN-E)', *Cryogenics (UK)*, 1990, **30**(Suppl.), pp. 794–798

97. Hull, J.R., Schoenung, S.M., Palmer, D.H. and Davis, M.K.: *Design and Fabrication Issues for Small-Scale SMES* (Argonne National Lab., Argonne, IL, USA, 1991)
98. Hassenzahl, W.V., Schainker, R.B. and Peterson, T.M.: 'Superconducting energy storage'. *Modem Power Systems (UK)*, 1991, **11**(3), pp. 27, 29, 31
99. Ter-Gazarian, A.G. and Zjebit, V.A.: 'Underground SMES construction particulars'. *Electricheskie Stantcii*, 1987, **9**, pp. 67–69
100. Tam, K.S. and Kumar, P.: 'Impact of superconductive magnetic energy storage on electric power transmission'. *IEEE Transactions on Energy Conversion*, 1990, **5**(3), pp. 501–511
101. Andrianov, V.V., Bashkirov, YU.A. and Ter-Gazarian, A.G.: 'Superconductor devices for the transmission, conversion and storage of energy in electric system'. *International Journal of High Temperature Superconductivity (Moscow)*, 1991, **2**
102. Astahov, YU.N., Venikov, V.A., Ter-Gazarian, A.G., Zhimerin, D.G. and Mohov, V.B.: 'Superconductive cable transmission line'. Author's Certificate No. 986 220, USSR, Priority from 17 July 1978
103. Astahov, YU.N., Venikov, V.A., Ter-Gazarian, A.G. and Zhimerin, D.G.: 'Superconductive cable line'. Author's Certificate No. 986 221, USSR, Priority from 17 July 1978
104. Astahov, YU.N., Venikov, V.A., Ter-Gazarian, A.G. and Zhimerin, D.G.: 'Superconductive DC transmission line'. Author's Certificate No. 743 465, USSR, Priority from 12 January 1979
105. Astahov, YU.N., Venikov, V.A., Ter-Gazarian, A.G. and Neporozjny, P.S.: 'Hydroaccumulator'. Author's Certificate No. 810 884, USSR, Priority from 31 August 1979
106. Astahov, YU.N., Venikov, V.A., Ter-Gazarian, A.G. and Sumin, A.G.: 'Power transmission line'. Author's Certificate No. 1 243 580, USSR, Priority from 9 December 1983
107. Astahov, YU.N., Venikov, V.A., Ter-Gazarian, A.G., Buhavtseva, N.A., Chigirev, A.I. and Yarnyh, L.V.: 'Means for nonsynchronous controllable connection between power systems'. Author's Certificate No. 1 305 173, USSR, Priority from 1 April 1985
108. Rosati, R.W., Peterson, J.L. and Vivirto, J.R.: *AC/DC Power Convertor for Batteries and Fuel Cells*, EPRI report EM-1286, Palo Alto, CA, USA, 1979
109. Shepherd, W. and Zand, P.: *Energy Flow and Power Factor in Nonsinusoidal Circuits* (Cambridge University Press, Cambridge, UK, 1979)
110. Ter-Gazarian, A.G. and Martynov, I.B.: 'Reactive power compensation and harmonics in power system with energy storage'. *Proceedings of Novosibirsk Electrotechnical Institute (Novosibirsk)*, 1985 (in Russian)
111. Ter-Gazarian, A.G. and Milyh, V.I.: 'AC/DC/AC control with the aim of energy losses decrease in power system'. *Proceedings of the Moscow Power Engineering Institute (Moscow)*, 1987, **104** (in Russian)
112. *An Assessment of Energy Storage Systems Suitable for Use by Electric Utilities*, EPRI/ERDA report EM-264, vol. 1, 1976

113. Barinov, V.A., Gurov, A.A., Korchak, V.U., Maneviteh, A.S. and Mitin, U. V.: 'Supply of consumers having a sharply varying load from power system'. *Electrichestvo*, 1990, **1**, pp. 1–5
114. Astahov, YU.N., Venikov, V.A. and Ter-Gazarian, A.G.: *Energy Storage in Power Systems* (Higher School, Moscow, 1989) (in Russian)
115. Infield, D.G.: *A Study of Electricity Storage and Central Electricity Generation*. SERC Rutherford Appleton Laboratory report RAL-84-045, 1984
116. Astahov, YU.N., Venikov, V.A., Ter-Gazarian, A.G., Kolobaev, P.B. and Sumin, A.G.: 'Energy storage parameters for nuclear power plants'. *Proceedings of the Moscow Power Engineering Institute (Moscow)*, 1983, **609**, pp. 77–81 (in Russian)
117. Astahov, YU.N., Venikov, V.A., Ter-Gazarian, A.G. and Sumin, A.G.: 'Energy storage influence on power system efficiency'. *Proceedings of the Novosibirsk Electrotechnical Institute (Novosibirsk)*, 1984 (in Russian)
118. Astahov, YU.N., Ter-Gazarian, A.G. and Cheremisin, I.M.: 'Pareto-optimisation in the conditions of uncertainty in some power system problems'. *Proceedings of the Moscow Agricultural Production Institute (Moscow)*, 1984 (in Russian)
119. Astahov, YU.N., Ter-Gazarian, A.G. and Sumin, A.G.: 'Structure optimisation for power system including energy storage'. *Proceedings of the Moscow Power Engineering Institute (Moscow)*, 1985, **65** (in Russian)
120. Astahov, YU.N., Ter-Gazarian, A.G. and Burkovsky, A.E.: 'Energy storage and discrete power engineering'. *Electrichestvo*, 1988, **1**
121. Ter-Gazarian, A.G. and Kagan, N.: 'Design model for electrical distribution systems considering renewable, conventional and energy storage sources'. *IEEE Proceedings (IEEE)*, 1992, **C-139**(6)
122. Astahov, YU.N., Venikov, V.A., Ter-Gazarian, A.G., Lidorenko, N.S., Muchnik, G.F., Ivanov, A.M., *et al.*: 'Energy storage functional capabilities in power system'. *Electrichestvo*, 1983, **4**, pp. 1–8
123. Astahov, YU.N. and Ter-Gazarian, A.G.: 'Throughput increase efficiency for transmission lines with the help of energy storage' in *Controllable Power Lines* (Shteentsa Publisher, Kishinev, 1986) (in Russian)
124. Kapoor, S.C.: 'Dynamic stability of long transmission systems with static compensators and synchronous machines'. *IEEE Transactions*, January/February 1979, **PAS-98**
125. Byerly, R.T., Poznaniak, E.R. and Taylor, E.R.: 'Static reactive compensation for power transmission systems'. *IEEE Transactions*, 1983, **PAS-102**, p. 3997
126. Aboytes, F., Arroyo, G. and Villa, G.: 'Application of static var-icompensators in longitudinal power systems'. *IEEE Transactions*, 1983, **PAS-102**, p. 3460
127. Rodgers, J.D., Schermer, R.I., Miller, B.L. and Hauer, J.I.: '30 MW SMES system for electric utility transmission stabilisation'. *Proceedings of IEEE*, 1983, **71**, pp. 1099–1108
128. McDanil, H.G. and Gabrielle, F.A.: 'Dispatching pumped storage generation'. *IEEE Transactions*, 1966, **PAS-85**, pp. 465–471

129. Farghal, S.A. and Shebble, K.M.: 'Management of power system generation applied to pumped storage plant', *Proceedings of the IEEE Canadian Communications and Energy Conference*, Montreal, Canada, October 1982, pp. 99–102
130. Lukic, V.P.: 'Optimal operating policy for energy storage'. *IEEE Transactions*, **PAS-101**, pp. 3295–3302
131. Ter-Gazarian, A.G. and Gladkov, V.G.: 'Criteria of optimal regimes for energy storage units'. *Proceedings of the Moscow Power Engineering Institute*, vol. 345, 1989, pp. 45–56
132. Laughton, M.A. (ed.): *Renewable Energy Sources*, Watt Committee on Energy report 22, 1990
133. Freris, L.L. (ed.): *Wind Energy Conversion Systems* (Prentice-Hall, Hertfordshire, UK, 1990)
134. Whittle, G.E.: 'Effects of wind power and pumped storage in an electricity generating system'. *Proceedings of the 3rd BWEA Conference (BWEA)*, 1981
135. Ter-Gazarian, A.G. and Watson, S.J.: 'The optimisation of renewable energy sources in an electrical power system by use of simulation and deterministic planning models'. *International Transactions in Operational Research*, October 1996, **3**(4), pp. 255–259
136. Ter-Gazarian, A.G. and Watson, S.J.: 'The operational and economic impact of renewable energy sources on three European Power Systems'. *Proceedings of the European Union Wind Energy Conference*, Göteburg, Sweden, May 1996
137. Ter-Gazarian, A.G., Watson, S.J., Halliday, J.A. and Davis, S.C.: 'Assessing the impact of grid integrated renewable energy sources on the economics of an expanding power system'. *Proceedings of the 15th British Wind Energy Association Conference*, York University, MEP Publications Ltd, London, 6–8 October 1993, pp. 139–145
138. Leicester, R.J., Newman, V.G. and Wright, L.K.: 'Renewable energy sources and storage'. *Nature*, 1978, **272**, pp. 518–521
139. Swift-Hook, D.T.: 'Firm power from the wind' in Galt J.M. (ed.), *Wind Energy Conversion* (MEP, London, 1987), p. 33
140. Swift-Hook, D.T.: 'How wind statistics affect the economic development of wind power' in H. K. Stephens (ed.),' *Proceedings of the ECWEC* (Herning, Bedford, UK, 1988), pp. 1022–1026
141. Ter-Gazarian, A.G. and Swift-Hook, D.T.: 'The value of storage on power systems with intermittent energy sources'. *Renewable Energy*, 1994, **5**(II), pp. 1479–1482
142. Swift-Hook, D.T.: 'Electricity from nuclear energy'. *Energy World*, April 1994, **217**, pp. 6–11
143. Swift-Hook, D.T.: 'Why renewable energy?' World Renewable Energy Congress X, WREC, Glasgow, 19–25 July 2008
144. Swift-Hook, D.T.: 'Intermittent generation as negative load'. (with D. J. Milborrow) World Renewable Energy Congress IX, WREC, Florence, 2006

Index